Practical Chromatography

Practical Chromatography

Jose Barnett

Larsen & Keller
www.larsen-keller.com

Practical Chromatography
Jose Barnett
ISBN: 978-1-64172-630-6 (Hardback)

⊟ Larsen & Keller

Published by Larsen and Keller Education,
5 Penn Plaza,
19th Floor,
New York, NY 10001, USA

Cataloging-in-Publication Data

Practical chromatography / Jose Barnett.
    p. cm.
Includes bibliographical references and index.
ISBN 978-1-64172-630-6
1. Chromatographic analysis. 2. Chemistry, Analytic. I. Barnett, Jose.
QD79.C4 P73 2022
543.8--dc23

For more information regarding Larsen and Keller Education and its products, please visit the publisher's website www.larsen-keller.com

TABLE OF CONTENTS

PREFACE

This book is a culmination of my many years of practice in this field. I attribute the success of this book to my support group. I would like to thank my parents who have showered me with unconditional love and support and my peers and professors for their constant guidance.

The laboratory technique used for the separation of a mixture is known as chromatography. The fluid in which the mixture is dissolved is called the mobile phase. It carries the fluid to another material named as the stationary phase. Various components of the mixture travel at different speeds due to which they separate. Chromatography can be either preparative or analytical. Preparative chromatography is concerned with the separation of a mixture's components for later use. Analytical chromatography deals with small amounts of material and is used to establish the presence and measurement of the relative proportions of analytes in a mixture. This book elucidates the concepts and innovative models around prospective developments with respect to chromatography. Some of the diverse topics covered herein address the varied techniques that fall under this category. This book will serve as a valuable source of reference for those interested in this area.

The details of chapters are provided below for a progressive learning:

Chapter – Introduction

The laboratory technique which is used for the separation of a mixture is referred to as chromatography. There are various types of chromatography including paper chromatography, thin layer chromatography, high performance liquid chromatography, gas chromatography, affinity chromatography, etc. This is an introductory chapter which will introduce briefly all the significant aspects of these types of chromatography.

Chapter – Gas Chromatography

Gas chromatography refers to a common type of chromatography which is used to separate and analyze compounds which can be vaporized without decomposition. This chapter has been carefully written to provide an easy understanding of gas chromatography including its principles, physical components, disadvantages and advantages.

Chapter – Liquid Chromatography

The analytical chromatographic technique that is used to separate ions and molecules that are dissolved in a liquid mobile phase is referred to as liquid chromatography. The usage of high pressure for separation, detection and quantification of different components in a mixture is termed as high performance liquid chromatography. The topics elaborated in this chapter will help in gaining a better perspective about the types and applications of high performance liquid chromatography.

Chapter – Paper Chromatography

The analytical method which is used to separate colored chemicals and substances is known as paper chromatography. Some of its types are ascending chromatography, two-dimensional chromatography and descending chromatography. The diverse applications of paper chromatography as well as its advantages and disadvantages have been thoroughly discussed in this chapter.

Chapter – Thin Layer Chromatography

Thin-layer chromatography is a technique in chromatography that is used for the separation of non-volatile mixtures. This chapter has been carefully written to provide an easy understanding of the basic principle as well as various processes, applications, advantages and disadvantages of thin-layer chromatography.

Chapter – Techniques in Chromatography

Some of the special techniques in chromatography are reversed-phase chromatography, fast protein liquid chromatography, countercurrent chromatography, periodic counter-current chromatography, capillary electrochromatography, etc. The chapter closely examines these techniques of chromatography to provide an extensive understanding of the subject.

Chapter – Uses of Chromatography

Chromatography is applied in various areas such as forensic testing, food regulation, athlete testing, etc. Chromatographic techniques are also used as effective methods of blood purification. The topics elaborated in this chapter will help in gaining a better perspective about these uses of chromatography.

Jose Barnett

Introduction 1

- **Principle of Seperation**
- **Different Components of Chromatography System**
- **Types of Chromatography**
- **Chromatogram**
- **Retention Factor**
- **Full Proof Techniques of Chromatography**
- **Chromatography Detector**

The laboratory technique which is used for the separation of a mixture is referred to as chromatography. There are various types of chromatography including paper chromatography, thin layer chromatography, high performance liquid chromatography, gas chromatography, affinity chromatography, etc. This is an introductory chapter which will introduce briefly all the significant aspects of these types of chromatography.

Chromatography is a technique for separating the components, or solutes, of a mixture on the basis of the relative amounts of each solute distributed between a moving fluid stream, called the mobile phase, and a contiguous stationary phase. The mobile phase may be either a liquid or a gas, while the stationary phase is either a solid or a liquid.

Kinetic molecular motion continuously exchanges solute molecules between the two phases. If, for a particular solute, the distribution favours the moving fluid, the molecules will spend most of their time migrating with the stream and will be transported away from other species whose molecules are retained longer by the stationary phase. For a given species, the ratio of the times spent in the moving and stationary regions is equal to the ratio of its concentrations in these regions, known as the partition coefficient. (The term adsorption isotherm is often used when a solid phase is involved.) A mixture of solutes is introduced into the system in a confined region or narrow zone (the origin), whereupon the different species are transported at different rates in the direction of fluid flow. The driving force for solute migration is the moving fluid, and the resistive force is the solute affinity for the stationary phase; the combination of these forces, as manipulated by the analyst, produces the separation.

Chromatography is one of several separation techniques defined as differential migration from a narrow initial zone. Electrophoresis is another member of this group. In this case, the driving force is an electric field, which exerts different forces on solutes of different ionic charge. The resistive force is the viscosity of the nonflowing solvent. The combination of these forces yields ion mobilities peculiar to each solute.

Chromatography has numerous applications in biological and chemical fields. It is widely used in biochemical research for the separation and identification of chemical compounds of biological origin. In the petroleum industry the technique is employed to analyze complex mixtures of hydrocarbons.

As a separation method, chromatography has a number of advantages over older techniques—crystallization, solvent extraction, and distillation, for example. It is capable of separating all the components of a multicomponent chemical mixture without requiring an extensive foreknowledge of the identity, number, or relative amounts of the substances present. It is versatile in that it can deal with molecular species ranging in size from viruses composed of millions of atoms to the smallest of all molecules—hydrogen—which contains only two; furthermore, it can be used with large or small amounts of material. Some forms of chromatography can detect substances present at the attogram (10^{-18} gram) level, thus making the method a superb trace analytical technique extensively used in the detection of chlorinated pesticides in biological materials and the environment, in forensic science, and in the detection of both therapeutic and abused drugs. Its resolving power is unequaled among separation methods.

Methods

Chromatographic methods are classified according to the following criteria: (1) geometry of the system, (2) mode of operation, (3) retention mechanism, and (4) phases involved.

Geometry

Column Chromatography

The mobile and stationary phases of chromatographic systems are arranged in such a way that migration is along a coordinate much longer than its width. There are two basic geometries: columnar and planar. In column chromatography the stationary phase is contained in a tube called the column. A packed column contains particles that either constitute or support the stationary phase, and the mobile phase flows through the channels of the interstitial spaces. Theory has shown that performance is enhanced if very small particles are used, which simultaneously ensures the additional desired feature that these channels be very narrow. The effect of mobile-phase mass transfer on band (peak) broadening will then be reduced. Constructing the stationary phase as a thin layer or film will reduce band broadening due to stationary-phase mass transfer. Porous particles, either as adsorbents or as supports for liquids, may have deep pores, with some extending through the entire particle. This contributes to band broadening. Use of microparticles alleviates this because the channels are shortened.

An alternate packing method is to coat impermeable macroparticles, such as glass beads, with a thin layer of microparticles. These are the porous-layer, superficially porous, or pellicular packings. As the particle size is reduced, however, the diameter of the column must also be decreased. As a result, the amount of stationary phase is less and the sample size must be reduced. Detection methods must therefore respond to very small amounts of solutes, and large pressures are required to force the mobile phase through the column. The extreme cases are known as microbore columns; an example is a column 35 centimetres (14 inches) long of 320-micrometre (1 micrometre = 10^{-4} centimetre) inside diameter packed with particles of 2-micrometre diameter.

A second column geometry involves coating the stationary phase onto the inside wall of a small-diameter stainless steel or fused silica tube. These are open tubular columns. The coating may be a liquid or a solid. For gaseous mobile phases, the superior performance is due to the length and the thin film of the stationary phase. The columns are highly permeable to gases and do not require excessive driving pressures. Columns in which a liquid mobile phase is used are much shorter and require large driving pressures.

Planar Chromatography

In this geometry the stationary phase is configured as a thin two-dimensional sheet. In paper chromatography a sheet or a narrow strip of paper serves as the stationary phase. In thin-layer chromatography a thin film of a stationary phase of solid particles bound together for mechanical strength with a binder, such as calcium sulfate, is coated on a glass plate or plastic sheet. One edge of the sheet is dipped in a reservoir of the mobile phase, which, driven by capillary action, moves through the bed perpendicular to the surface of the mobile phase. This capillary motion is rapid compared to solute diffusion in the mobile phase at right angles to the migration path, and so the solute is confined to a narrow path.

Mode of Operation

Development Chromatography

In terms of operation, in development chromatography the mobile phase flow is stopped before solutes reach the end of the bed of stationary phase. The mobile phase is called the developer, and the movement of the liquid along the bed is referred to as development. With glass columns of diameter in the centimetre range and large samples (cubic-centimetre range), the bed is extruded from the column, the solute zones carved out, and solutes recovered by solvent extraction. Although this is easily done with coloured solutes, colourless solutes require some manner of detection, such as ultraviolet light absorption or fluorescence or the streaking of the column with a reagent that reacts with the solute to form a coloured product.

Planar systems involve placing the samples (in the 10^{-3} cubic-centimetre range) as spots at an edge of the stationary bed parallel to the developer. Solute zones are located by light irradiation or by spraying the bed with a colour-producing reagent. Migration is reported in terms of the *Rf* value, the distance moved by the centre of the zone relative to the distance moved by the mobile phase front, where both are measured from the origin. Use of the solvent front as a reference point is frequently inconvenient. A standard solute is often included, and the migration of the solutes relative to the standard reported as the relative R value. If larger samples are required for subsequent manipulation, either simultaneous separations are performed or the sample is applied as a streak across the stationary phase. The final spot or band is carved or cut from the chromatogram. In one type of planar chromatography, the mixture is placed at one corner of a square bed, plate, or sheet and developed, the mobile phase is evaporated, and the plate is rotated 90° so that the spots become the origins for a second development with a different developer. This is termed two-dimensional planar chromatography.

Elution Chromatography

This method, employed with columns, involves solute migration through the entire system and solute detection as it emerges from the column. The detector continuously monitors the amount of solute in the emerging mobile-phase stream—the eluate—and transduces the signal, most often to a voltage, which is registered as a peak on a strip-chart recorder. The recorder trace where solute is absent is the baseline. A plot of the solute concentration along the migration coordinate of development chromatograms yields a similar solute peak. Collectively the plots are the concentration profiles; ideally they are Gaussian (normal, bell, or error curves). The signal intensity may also be digitized and stored in a computer memory for recall later. Solute behaviour is reported in terms of the retention time, which is the time required for a solute to migrate, or elute, from the column, measured from the instant the sample is injected into the mobile phase stream to the point at which the peak maximum occurs. The adjusted retention time is measured from the appearance of an unretained solute at the outlet. The dependence of these times on flow rate is removed by reporting the retention volumes, which are calculated as the retention times multiplied by the volumetric flow rate of the mobile phase.

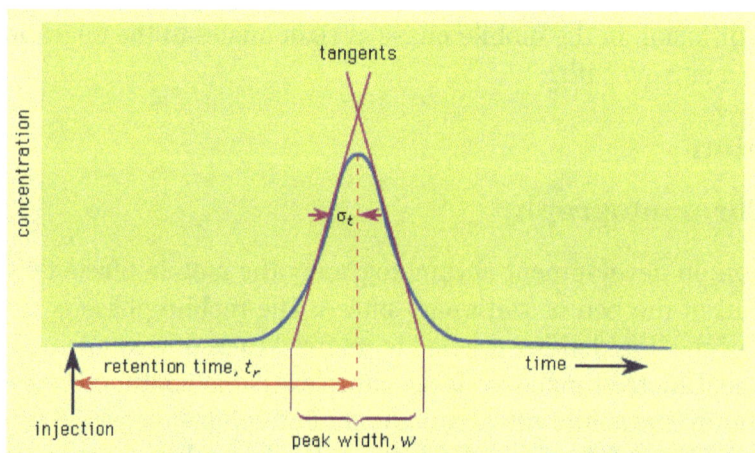

Peak shape, peak width, and plate height parameters in elution chromatography.

The spots on the developed planar bed, the series of peaks on the paper produced by the recorder, or the printout of the computer data are various forms of chromatograms.

Retention Mechanism

Classification in terms of the retention mechanism is approximate, because the retention actually is a mixture of mechanisms. If the partition coefficient is constant as the amount of solute is varied, the separation is referred to as linear chromatography. This condition is highly desirable because solute zones approach symmetrical Gaussian distributions. If the system is nonlinear, solute zones are asymmetrical. In the most common asymmetrical case, a zone "tails" into a following solute zone to contaminate it.

In adsorption chromatography solute molecules bond directly to the surface of the stationary phase. Stationary phases may contain a variety of adsorption sites differing in the tenacity with which they bind the molecules and in their relative abundance. The net effect determines the adsorbent activity. Partition chromatography utilizes a support material coated with a stationary-phase liquid. Examples are (1) water held by cellulose, paper, or silica, or (2) a thin film coated or bonded to a solid. The solid support ideally is inactive in the retention of solutes, but it actually is not; retention is mostly due to solute solution in the stationary liquid phase.

The stationary phase in size-exclusion chromatography consists of molecules of the mobile phase trapped in the porous structure of a solid. Solute molecules are retained when they diffuse into and out of these pores. The time they remain in the pores is a function of their size, which determines the depth of penetration. There is a certain molecular size that represents the "just excluded" case. Molecules of this size and larger are excluded from the pores and are not separated. They appear first in elution chromatography. At the other end of the size spectrum, there is a certain size for which all molecules of this magnitude and smaller penetrate all the pores. These molecules also are not separated; they elute last. Gel-filtration chromatography refers to size-exclusion methods employing water as the mobile phase; gel-permeation chromatography makes use of an organic mobile phase.

Very specific intermolecular interactions, "lock and key," are known in biochemistry. Examples include enzyme-protein, antigen-antibody, and hormone-receptor binding. A structural feature of an enzyme will attach to a specific structural feature of a protein. Affinity chromatography exploits this feature by binding a ligand with the desired interactive capability to a support such as a gel used in gel-filtration chromatography. The ligand retards a solute with the compatible structural feature and passes all other solutes in the mixture. The solute is then eluted by a mobile-phase change such as incorporating a competing solute, changing the acidity, or changing the ionic strength of the eluent.

There is no stationary phase in field-flow fractionation; the different-velocity streams or layers of the mobile phase with the solute distributed between them produce the separation.

Phases

Gas Chromatography

Classification by phases gives the physical state of the mobile phase followed by the state of the stationary phase. Gas chromatography employing a gaseous fluid as the mobile phase, called the carrier gas, is subdivided into gas-solid chromatography and gas-liquid chromatography. The carrier gases used, such as helium, hydrogen, and nitrogen, have very weak intermolecular interactions with solutes. Molecular sieves are used in gas size-exclusion chromatography applied to gases of low molecular weight. Adsorption on solids tends to give nonlinear systems. Gas-liquid chromatography employs a liquid stationary phase where solution forces provide retention. At ordinary pressures the solutes in the gas phase behave as a mixture of ideal gases. All interactions responsible for selective retention occur in the stationary phase. Thus, a wide variety of liquid stationary phases have been employed; hundreds have been reported.

A basic rule in organic chemistry is that "like dissolves like." Thus, the polar solvent water dissolves the polar solute ethanol but not the hydrocarbon octane. The nonpolar solvent benzene will dissolve octane but not ethanol. Polar stationary phases will retain polar solutes and pass those that are nonpolar. The order of emergence is reversed with nonpolar stationary phases. Lutz Rohrschneider of Germany initiated studies that led to a standard set of solute species, solvent probes, which helped order stationary phases in terms of polarity and intermolecular interactions present.

In gas chromatography the retention of solutes is most often referred to the behaviour of the straight-chain hydrocarbons; i.e., relative retention volumes are used. On a logarithmic scale this becomes the retention index (RI) introduced by the Swiss chemist Ervin sz. Kováts. The RI values of the solvent probes serve as the basis for the classification method introduced by Rohrschneider. Similar schemes have been suggested for liquid systems.

Gas-phase intermolecular interactions occur and are exploited in supercritical-fluid chromatography. Examples of interactive gases used at high pressure are carbon dioxide, nitrous oxide, ammonia, hydrocarbons, sulfur hexafluoride, and halogenated methanes.

Mixtures of solutes that have a wide boiling point or polarity range or have a large variety of functional groups pose a particular problem. At low column-operating temperatures, the solutes with high volatility (or, more precisely, solutes with a large numerical value for the liquid solution activity coefficient) appear early on the chromatogram as well-resolved peaks. Solutes with low volatility progress slowly through the column, with ample opportunity for the peak broadening. These solutes appear as very low, broad peaks that may be overlooked. An increase in column temperature increases the concentration of the solutes in the gas phase. The solutes of high volatility, however, now spending most of their time in the mobile-gas phase, migrate rapidly through the column to appear as unresolved peaks. The succeeding solutes are adequately resolved. This is termed the general elution problem. A simple solution is to increase the column temperature during the course of the separation. The well-resolved, highly volatile solutes are removed from the column at the lower temperatures before the low-volatility solutes leave the origin at the column inlet. This technique is termed temperature-programmed gas chromatography.

Liquid Chromatography

This form of chromatography employs a liquid mobile phase. Liquid-solid chromatography utilizes a solid stationary phase, and the major mechanism of retention is adsorption. Popular adsorbents are silica and alumina, which both retain polar compounds. If a polar mobile phase is used, the solutes are rapidly swept from the bed. Thus, the preferred mobile phase is a nonpolar or slightly polar solvent. The American chemist Lloyd R. Snyder arranged solvents in an eluotropic strength scale based on the chromatographic behaviour of selected solutes on silica. Normal-phase chromatography involves a polar stationary phase and a less polar mobile phase.

Liquid-liquid chromatography employs liquid mobile and stationary phases. High-performance liquid chromatography uses small particles with molecules bonded to their surface to give a thin film that has liquidlike properties. A number of bonding agents are available. A nonpolar molecule can be bonded to the solid and a polar mobile phase used. This method is termed reverse-phase liquid chromatography. The partition coefficient depends on the identity of both mobile and stationary phases. In this case, however, the number of stationary phases is limited, while there is a large number of liquids and combinations of them used for the mobile phase. Mobile phases of constant composition are called isocratic.

The general elution problem encountered in liquid chromatography involves samples that contain both weakly and strongly retained solvents. This is handled in a manner analogous to the temperature programming used in gas chromatography. In a process termed gradient elution, the concentration of well-retained solutes in the mobile phase is increased by constantly changing the composition, and hence the polarity, of the mobile phase during the separation.

Sample Recovery

Sample recovery from development chromatograms has been described—that is, detection followed by carving zones from an extruded column or carving or cutting zones from the planar stationary-phase bed. In elution chromatography successive samples of the effluent are collected in tubes held in a mechanically driven rotating tray called a fraction collector. Analogous arrangements exist to condense and trap solutes from effluent gas streams. Large samples can be used to prepare relatively large amounts of pure solutes for further manipulation; this is the realm of preparative-scale chromatography.

Methods of Detection

High-resolution gas or liquid elution chromatography of multicomponent samples deals with small amounts of solutes emerging from the column where they are to be detected. Refinement of chromatographic methods is inseparable from refinement of detectors that accurately sense solutes in the presence of the mobile phase. Detectors may be classified as general detectors in which all solutes are sensed regardless of their identity, or as specific detectors, which sense a limited number of solutes—for example, those containing halogens or nitrogen. Detectors may be nondestructive, whereby sensing does not alter the nature of the solutes, as in the case of light absorption, so they may be collected for further use. Destructive detectors, on the other hand, destroy the solutes. Detectors include not only the component that senses the solutes but also those that perform the associated transduction, electronic amplification, and final readout.

Detector Characteristics

There are three essential detector characteristics. The first is the lower limit of detection, the smallest amount of solute measured in terms of moles (mass-sensitive detectors) or moles per litre (concentration-sensitive detectors) that can be detected; this entails distinguishing a signal from the random noise inherent in all electronic systems. A second is the sensitivity, which is the change in signal intensity per unit change in the amount of solute. The third is the linear range—i.e., the range of solute amount where the signal intensity is directly proportional to the amount of solute; doubling the amount doubles the signal intensity. Solutes may respond differently to a detector. For example, if equal amounts of methane (containing one carbon) and ethane (two carbons) enter a flame-ionization detector, the peak for ethane will be twice the size of that for methane. The detector acts as a "carbon counter." A response factor may be determined for each solute to accommodate this. The perfect detector ideally has "zero volume"; that is, only an infinitesimal amount of solute enters the sensing region, produces a signal, and exits before the next infinitesimal amount enters the detector chamber. In the worst case, a solute enters the detector chamber and remains there producing a signal while the next portion of the solute enters behind it. This invites the possibility of a solute still being present and producing a signal as a succeeding solute of a different kind enters the sensing region, thereby undoing the separation achieved by the column. In addition, the readout system should have an instantaneous response time. Mechanical systems such as strip-chart recorders have an inertia, so that if an electrical pulse enters the circuit a small but finite time is required for the recorder pen to reach its final position. The dead-band is that region of the signal in which the system does not respond to small changes in the amount of solute; there is "slack" in the system. Such imprecision becomes insignificant, however, if sample injection is not instantaneous. The injected sample must not reside in a prechamber that slowly feeds it onto the column. The chromatogram should report everything that happens, from sample injection to the final data presentation. The most challenging detection problem is a sample containing a wide variety of solutes that covers a large range of concentrations and produces very closely spaced, narrow peaks.

Gas Chromatographic Detectors

Gas chromatographic detectors sense the solute vapours in the mobile phase as they emerge from the column. Thermal-conductivity detectors compare the heat-conducting ability of the exit gas stream to that of a reference stream of pure carrier gas. To accomplish this, the gas streams are passed over heated filaments in thermal-conductivity cells. Measured changes in filament resistance of the cells reflect temperature changes caused by increments in thermal conductivity. This resistance change is monitored and registered continuously on a recorder. An alternate type of detector is the flame-ionization detector, in which the gas stream is mixed with hydrogen and burned. Positive ions and electrons are produced in the flame when organic substances are present. The ions are collected at electrodes and produce a small, measurable current. The flame-ionization detector is highly sensitive to hydrocarbons, but it will not detect carrier gases, such as nitrogen, or highly oxidized materials, such as carbon dioxide, carbon monoxide, sulfur dioxide, and water. In another device, the electron-capture detector, a stream of electrons from a radioactive source is produced in a potential field. Materials in the gas stream containing atoms of certain types capture electrons from the stream and measurably reduce the current. The most important of the capturing atoms are the halogens—fluorine, chlorine, bromine, and iodine. This type of detector, therefore, is particularly useful with chlorinated pesticides. Certain elements will emit light of distinctive wavelength when excited in a flame. The flame photometric detector measures the intensity of light with a photometric circuit. Solute species containing halogens, sulfur, or phosphorus can be burned to produce ionic species containing these elements and the ions sensed by electrochemical means.

Liquid Chromatographic Detectors

Liquid chromatographic detectors suitable for high-performance columns require clever technology. If the solutes contain structural features that absorb light at certain wavelengths, the decrease in the intensity of the transmitted beam of light compared to the intensity of the incident beam can be used to monitor the effluent stream. In order for the solute to be detected, it must contain light-absorbing groups, the excitation source must contain light of a wavelength peculiar to this group, and the photoelectric sensor must respond to this wavelength. Also, the mobile phase must be transparent at this wavelength. The scope of solute species detected can be enlarged by reacting a light-insensitive solute with a reagent that contains a light-sensitive group and passing the product through the detector. Solutes may contain groups that absorb light at one wavelength and reemit light of a different wavelength. The fluorescence detector responds to these substances. Light bends or refracts on passing through an interface between air and a liquid or liquid solution. The degree of refraction depends on the nature of the liquid or the composition of the solution. The refractive index detector compares the refraction of the pure mobile phase with that of the column effluent.

Chromatography–Mass Spectrometry Methods

The mass spectrometer is an analytical instrument that bombards molecules with a stream of electrons in a chamber at extremely low pressure to produce a stream of charged fragments that differ in mass. The population of the fragments and the ratio of mass to charge is characteristic of the target molecule. Each fragment is deflected differently in a magnetic field to produce a pattern, the mass spectrum, which can be used to identify the target. The system is a very specific identifying detector when coupled with chromatography. The spectrum can be stored in a computer and compared with entries in a mass spectrum library. For some time the problem with gaseous effluents had been to match the column effluent at one atmosphere pressure to the high-vacuum inlet of the mass spectrometer, while the problem with liquid chromatography had been the large amount of mobile phase entering the ionization chamber of the spectrometer. These incompatibility problems have finally been overcome, and the mass spectrometer is now used in both gas and liquid chromatography. The technology of mass spectrometry is as great, if not greater, than that of chromatography.

If mass spectral data are lacking, solutes in a sample are identified by comparing their behaviour with that of known compounds. In gas chromatography this is best done by determining the retention index of the unknown solute and comparing it with the tabulated data for known compounds on the stationary phase used. Methods exist for estimating the effect of temperature and temperature programming on the retention index.

The area enclosed by a peak, suitably adjusted for the detector response factor for that solute, is proportional to the amount of solute producing the peak. The area is frequently approximated from the peak width and height. Modern electronic integrators will, when properly instructed, ignore electronic noise, compensate for baseline drift, start integration when a peak appears, integrate, and stop the process when the peak exits the detector. Integration, a process of summation, is accomplished by opening and closing a narrow electronic window, registering the signal intensity, repeating the process, and then summing the stored signals to produce a number proportional to the area. The integrator will also sense the peak maximum. The chromatogram is a printed tape with the retention times and peak areas. Programs exist that will incorporate the response factor and calculate the relative peak areas, which give the percentage composition of the sample. Stored mass spectral data may be manipulated to produce the same data. Peak heights are used as quantitative measures for narrow peaks for which the area is difficult to determine accurately.

Efficiency and Resolution

There are two features of the concentration profile important in determining the efficiency of a column and its subsequent ability to separate or resolve solute zones. Peak maximum, the first, refers to the location of the maximum concentration of a peak. To achieve satisfactory resolution, the maxima of two adjacent peaks must be disengaged. Such disengagement depends on the identity of the solute and the selectivity of the stationary and mobile phases.

The second feature important to efficiency and resolution is the width of the peak. Peaks in which the maxima are widely disengaged still may be so broad that the solutes are incompletely resolved. For this reason, peak width is of major concern in chromatography.

Column Efficiency

The efficiency of a column is reported as the number of theoretical plates (plate number), N, a concept Martin borrowed from his experience with fractional distillation:

$$N = 16\left(\frac{t_r}{w}\right)^2,$$

where t_r is the retention time measured from the instant of injection and w is the peak width obtained by drawing tangents to the sides of the Gaussian curve at the inflection points and extrapolating the tangents to intercept the baseline. The distance between the intercepts is the peak width. If the peak is a Gaussian distribution, statistical methods show that its width may be determined from the standard deviation, σ, by the formula $w = 4\sigma$. Poor chromatograms are those with early peaks (small t_r) that are broad (large w), hence giving small N values, while excellent chromatograms are those with late-appearing peaks (large t_r) that are still very narrow (small w), thereby producing a large N. The number of theoretical plates is a measure of the "goodness" of the column. Plate numbers may range from 100 to 10^6. The peak width determined from the chromatogram includes contributions from the sample-injection technique, extraneous tubing, and the detector. These are extra column contributions to peak broadening. Although very important, they are not part of the chromatographic process and will be ignored here. The plate number depends on the length of the column.

The extreme value of 10^6 plates was obtained with an open tubular gas chromatographic column 1.6 kilometres (1 mile) long. A more appropriate parameter for measuring efficiency is the height equivalent to a theoretical plate (or plate height), HETP (or h), which is L/N, L being the length of the column. Efficient columns have small h values.

Resolution

In general, resolution is the ability to separate two signals. In terms of chromatography, this is the ability to separate two peaks. Resolution, R, is given by,

$$R = \frac{\left(t_{r2} - t_{r1}\right)}{1/2\left(w_1 + w_2\right)}.$$

where t_{r1} and t_{r2} and w_1 and w_2 are the times and widths, respectively, of the two immediately adjacent peaks. If the peaks are sufficiently close, which is the pertinent problem, w is nearly the same for both peaks and resolution may be expressed as,

$$R = \frac{\left(t_{r2} - t_{r1}\right)}{4\sigma}.$$

If the distance between the peaks is 4σ, then R is 1 and 2.5 percent of the area of the first peak overlaps 2.5 percent of the area of the second peak. A resolution of unity is minimal for quantitative analysis using peak areas.

Theoretical Considerations

Retention

The rates of migration of substances in chromatographic procedures depend on the relative affinity of the substances for the stationary and the mobile phases. Those solutes attracted more strongly to the stationary phase are held back relative to those solutes attracted more strongly to the mobile phase. The forces of attraction are usually selective—that is to say, stronger for one solute than another. At least one of the two phases must exert a selective effect, and very often both phases are selective, as in liquid and supercritical-fluid chromatography. In gas chromatography, the mobile phase is ordinarily a gas that exerts essentially no attractive force on the solutes at all. In this case, the mobile phase is entirely nonselective.

The forces attracting solutes to the two phases are the normal forces existing between molecules—intermolecular forces. There are five major classes of these forces: (1) the universal, but weak, interaction between all electrons in neighbouring atoms and molecules, called dispersion forces, (2) the induction effect, by which polar molecules (those having an asymmetrical distribution of electrons) bring about a charge asymmetry in other molecules, (3) an orientation effect, caused by the mutual attraction of polar molecules resulting from alignment of dipoles (positive charges separated from negative charges), (4) hydrogen bonding between dipolar molecules bearing electron-pair-accepting hydrogen atoms, and (5) acid-base interactions in the Lewis acid-base sense—i.e., the affinity of electron-accepting species (Lewis acids) to electron donors (Lewis bases). The interplay of these forces and temperature are reflected in the partition coefficient and determine the order on polarity and eluotropic strength scales. In the special case of ions, a strong electrostatic force exists in addition to the other forces; this electrostatic force attracts each ion to ions of opposite charge. This is an important element of ion-exchange chromatography.

Plate Height

In chromatography, peak width increases in proportion to the square root of the distance that the peak has migrated. Mathematically, this is equivalent to saying that the square of the standard deviation is equal to a constant times the distance traveled. The height equivalent to a theoretical plate, as discussed above, is defined as the proportionality constant relating the standard deviation and the distance traveled. Thus, the defining equation of the height equivalent to a theoretical plate is as follows: HETP = σ^2/L, in which σ is the standard deviation and L the distance traveled. The use of the plate height is superior to the use of peak width in evaluating various chromatographic systems, because it is constant for the chromatographic run, and it is nearly constant from solute to solute.

In elution chromatography, in which the peak develops on a time scale, an equivalent form of the above equation is HETP = $L \sigma t^2/tr^2$, in which L is now the column length, tr the time of retention of the peak by the column, and σt the standard deviation of the peak measured in units of time; this form is another expression of the equation HETP = L/N given previous.

During a chromatographic separation, three basic processes contribute to plate height (HETP): (1) Molecular diffusion, in which solute molecules diffuse outward from the centre of the zone. This effect is inversely proportional to the average linear flow velocity, u, because rapid flow reduces the time for diffusion. Mathematically, the contribution to plate height of this factor is expressed as B/u, in which B is a constant. (2) Eddy diffusion, in which solute is carried at unequal rates through the tortuous pathways of the granular bed of the packing particles. The contribution to plate height is a constant factor, A, independent of velocity. (3) Nonequilibrium or mass transfer, in which the slowness of diffusion in and out of the stationary and mobile phases causes fluctuations in the times of residence of the solute in the two phases and a consequent peak broadening. The effect is proportional to velocity and is expressed as $C_s u$ and $C_m u$, in which C_s and C_m are constants relating to the stationary and mobile phases, respectively.

A function of chromatographic theory has been twofold: (1) to evaluate B, A, C_m, and C_s, in terms of underlying diffusivity and flow processes, and (2) to assemble them into a total plate height equation.

The general equation used is HETP = $A + B/u + C_s u$. This is inadequate at high velocities, however, and is replaced by the equation,

$$\text{HETP} = \left(\frac{1}{A} + \frac{1}{C_m u} \right)^{-1} + \frac{B}{u} + C_3 u.$$

Knowledge of the component terms in such equations allows one to optimize chromatographic operating conditions.

Principle of Seperation

The mixture of compound 1 and 3 is shown in figure and assume if we are using boiling point as a criteria to isolate them. As we will heat the mixture there will two phase forms, one liquid phase and other is vapor phase. The molecules of compound 1 and 3 will distribute between these two phases and as the temp is near to boiling point of compound 1, more amount of 1 will be present in vapor phase than liquid phase. Where as more number of compound 3 will be in liquid phase. Eventually as this process will continue, at the end two molecules will get separated from each other. The distribution coefficient (Kd) to describe the distribution of compound 1 between two phase A and B is as follows:

$$Kd = \frac{\text{Concentration in Phase A}}{\text{Concentration in Phase B}}$$

Similarly one can also exploit other physical & chemical parameters as well. With each and every physical and chemical parameter the molecule present in the mixture will distribute as per their behavior in each parameter.

Different Components of Chromatography System

The different components of a chromatography system are given in figure. It has following components:

1. Reservoir: One or two reservoir for mobile phase (buffer).

2. Pump: One or two pump to flow the buffer from reservoir. Different types of pumps are used in chromatography system, mostly based on the pressure level required to perform chromatography. A pump is chosen as per the pressure required to run the mobile phase. Based on the pressure level, liquid chromatography can be classified into the following categories:

    Low Pressure Liquid Chromatography: Pressure limit less than 5 Bar.

    Medium Pressure Liquid Chromatography: Intermediate pressure limit (6-50 bar).

    High Pressure Liquid Chromatography: Pressure limit more than 50-350 bar. A typical polysaccharide bead is not appropriate to withstand high pressure during HPLC. Hence, in HPLC silica based beads are recommended. Due to high pressure and smaller size of the silica beads gives higher number of theoretical plates. This gives HPLC superior resolving power to separate complex biological samples.

3. Mixer: A mixer is required to mix the buffer received from both pumps to form a linear or step gradient.

4. Column: A column made up of glass or steel.

5. Detector: The elution coming out from column goes to the online monitoring system to test the presence of the analyte based on different properties. There are different types of detectors are known in chromatography such as UV-Visible detector etc.

6. Fraction Collection: The eluent can be collected in different fractions by a fraction collector.

7. Recorder: The profile of eluent with respect to the measured property in a detector can be plotted in the recorder.

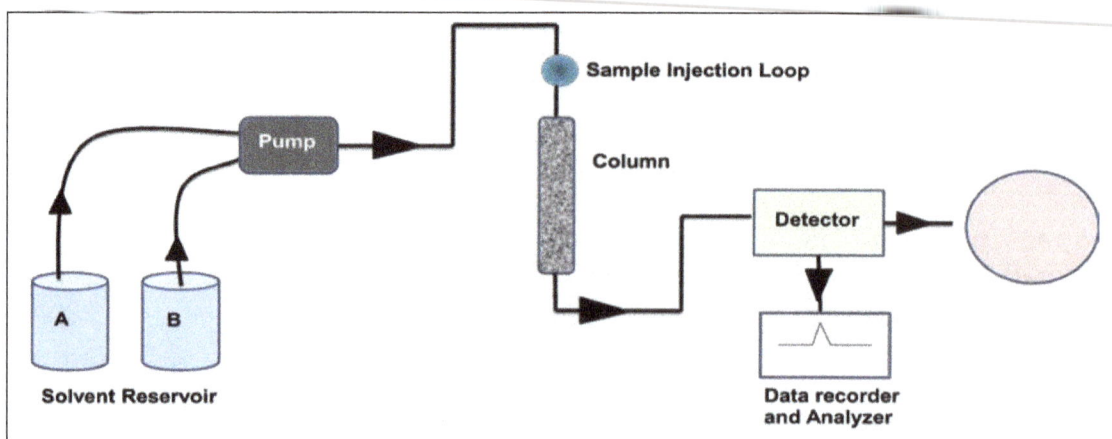

Different components of a chromatography system.

Types of Chromatography

There are different kinds of chromatographic techniques and these are classified according to the shape of bed, physical state of mobile phase, separation mechanisms.

Apart from these there are certain modified forms of these chromatographic techniques involving different mechanisms and are hence categorized as modified or specialized chromatographic techniques.

Column Chromatography

It is the preparative application of chromatography. It is used to obtain pure chemical compounds from a mixture of compounds on a scale from micrograms up to kilograms using large industrial columns. The classical preparative chromatography column is a glass tube with a diameter from 5 to 50 mm and a height of 50 cm to 1 m with a tap at the bottom.

A chemist using column chromatographic apparatus.

Slurry is prepared of the eluent with the stationary phase powder and then carefully poured into the column. Care must be taken to avoid air bubbles. A solution of the organic material is pipetted on top of the stationary phase.

This layer is usually topped with a small layer of sand or with cotton or glass wool to protect the shape of the organic layer from the velocity of newly added eluent. Eluent is slowly passed through the column to advance the organic material. Often a spherical eluent reservoir or an eluent-filled and stoppered separating funnel is put on top of the column.

The individual components are retained by the stationary phase differently and separate from each other while they are running at different speeds through the column with the eluent. At the end of the column they elute one at a time. During the entire chromatography process the eluent is collected in a series of fractions.

The composition of the eluent flow can be monitored and each fraction is analyzed for dissolved compounds, e.g., by analytical chromatography, UV absorption, or fluorescence. Coloured compounds (or fluorescent compounds with the aid of an UV lamp) can be seen through the glass wall as moving bands.

The stationary phase or adsorbent in column chromatography is a solid. The most common stationary phase for column chromatography is $-C_{18}H_{37}$, followed by alumina. Cellulose powder has often been used in the past. Also possible are ion exchange chromatography, reversed-phase chromatography (RP), affinity chromatography or expanded bed adsorption (EBA). The stationary phases are usually finely ground powders or gels and/or are micro porous for an increased surface; though in EBA a fluidized bed is used.

The mobile phase or eluent is either a pure solvent or a mixture of different solvents. It is chosen so that the retention factor value of the compound of interest is roughly around 0.75 in order to minimize the time and the amount of eluent to run the chromatography. The eluent has also been chosen so that the different compounds can be separated effectively. The eluent is optimized in small scale pretests, often using thin layer chromatography (TLC) with the same stationary phase.

A faster flow rate of the eluent minimizes the time required to run a column and thereby minimizes diffusion, resulting in a better separation. A simple laboratory column runs by gravity flow. The flow rate of such a column can be increased by extending the fresh eluent filled column above the top of the stationary phase or decreased by the tap controls. Better flow rates can be achieved by using a pump or by using compressed gas (e.g., air, nitrogen, or argon) to push the solvent through the column (flash column chromatography).

Automated flash chromatography systems attempt to minimize human involvement in the purification process. Automated systems may include components normally found on HPLC systems (gradient pump, sample injection apparatus, UV detector) and a fraction collector to collect the eluent. The software controlling an automated system will coordinate the components and help the user to find the resulting purified material within the fraction collector. The software will also store results from the process for archival or later recall purposes.

Paper Chromatography

It is an analytical technique for separating and identifying mixtures that are or can be coloured, especially pigments. This can also be used in secondary or primary schools in ink experiments. This method has been largely replaced by thin layer chromatography; however it is still a powerful teaching tool. Two-way paper chromatography, also called two-dimensional chromatography, involves using two solvents and rotating the paper 90° in between. This is useful for separating complex mixtures of similar compounds, for example, amino acids.

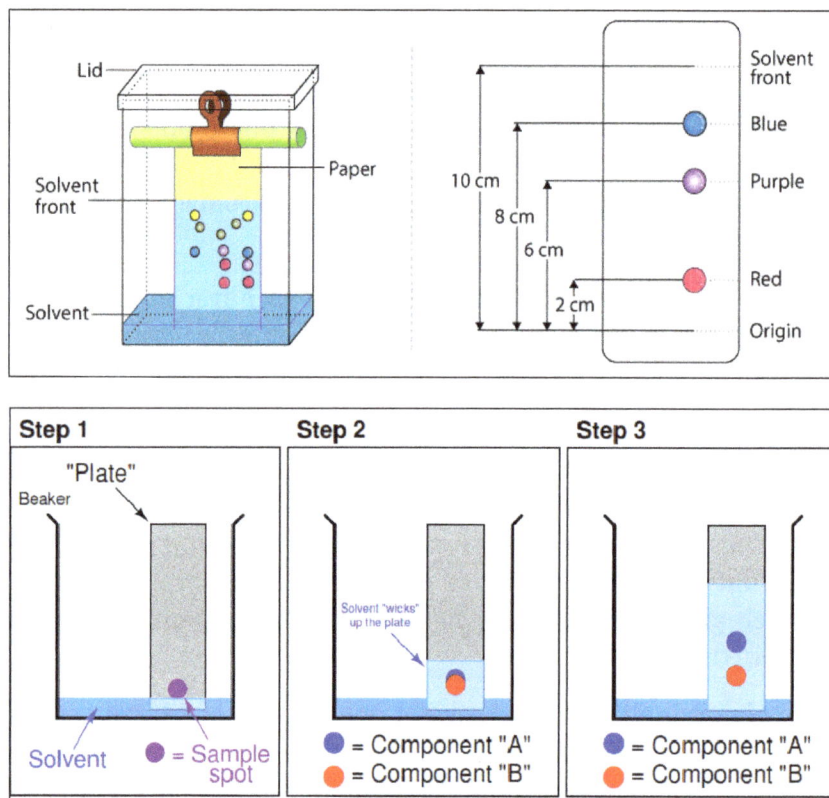

In some cases, paper chromatography does not separate pigments completely; this occurs when two substances appear to have the same values in a particular solvent. In these cases, two-way chromatography is used to separate the multiple-pigment spots. The chromatogram is turned by ninety degrees, and placed in a different solvent in the same way as before; some spots separate in the presence of more than one pigment.

As before, the value is calculated, and the two pigments are identified. The R_f value (retention factor) is the distance travelled by a particular component from the origin (where the sample was originally spotted) as a ratio to the distance travelled by the solvent front from the origin. Rf values for each substance will be unique, and can be used to identify components. A particular component will have the same R_f value if it is separated under identical conditions.

After development, the spots corresponding to different compounds may be located by their colour, ultraviolet light, ninhydrin (Triketohydrindane hydrate) or by treatment with iodine vapours. The final chromatogram can be compared with other known mixture chromatograms to identify sample mixture using the R_n value.

As in most other forms of chromatography, paper chromatography uses R_n values to help identify compounds. R_f values are calculated by dividing the distance the pigment travels up the paper by the distance the solvent travels (the solvent front). Because R_f values are standard for a given compound, known R_n values can be used to aid in the identification of an unknown substance in an experiment.

High Performance Liquid Chromatography

High-performance liquid chromatography (HPLC) is a form of column chromatography used frequently in biochemistry and analytical chemistry. It is also sometimes referred to as high-pressure liquid chromatography. HPLC is used to separate components of a mixture by using a variety of chemical interactions between the substance being analyzed (analyte) and the chromatography column.

In isocratic HPLC, the analyte is forced through a column of the stationary phase (usually a tube packed with small round particles with a certain surface chemistry) by pumping a liquid (mobile phase) at high pressure through the column. The sample to be analyzed is introduced in a small volume to the stream of mobile phase and is retarded by specific chemical or physical interactions with the stationary phase as it traverses the length of the column.

From left to right: A pumping device generating a gradient of two different solvent.
A steel enforced column and an apparatus for measuring the absorbance.

The amount of retardation depends on the nature of the analyte, stationary phase and mobile phase composition. The time at which a specific analyte elutes (comes out of the end of the column) is called the retention time and is considered a reasonably unique identifying characteristic of a given analyte. The use of pressure increases the linear velocity (speed) giving the components less time to diffuse within the column, leading to improved resolution in the resulting chromatogram.

Common solvents used include any miscible combinations of water or various organic liquids (the most common are methanol and acetonitrile). Water may contain buffers or salts to assist in the separation of the analyte components, or compounds such as Trifluoroacetic acid which acts as an ion pairing agent.

A further refinement to HPLC has been to vary the mobile phase composition during the analysis; this is known as gradient elution. A normal gradient for reverse phase chromatography might start at 5% methanol and progress linearly to 50% methanol over 25 minutes, depending on how hydrophobic the analyte is.

The gradient separates the analyte mixtures as a function of the affinity of the analyte for the current mobile phase composition relative to the stationary phase. This partitioning process is similar to that which occurs during a liquid-liquid extraction but is continuous, not step-wise. In this example, using a water/methanol gradient, the more hydrophobic components will elute (come off the column) under conditions of relatively high methanol; whereas the more hydrophilic compounds will elute under conditions of relatively low methanol.

The choice of solvents, additives and gradient, depends on the nature of the stationary phase and the analyte. Often a series of tests are performed on the analyte and a number of generic runs may be processed in order to find the optimum HPLC method for the analyte — the method which gives the best separation of peaks.

Types of HPLC

Normal Phase Chromatography

Normal phase HPLC (NP-HPLC) was the first kind of HPLC chemistry used, and separates analytes based on polarity. This method uses a polar stationary phase and a non-polar mobile phase, and is used when the analyte of interest is fairly polar in nature. The polar analyte associates with and is retained by the polar stationary phase.

Adsorption strengths increase with increase in analyte polarity, and the interaction between the polar analyte and the polar stationary phase (relative to the mobile phase) increases the elution time. The interaction strength not only depends on the functional groups in the analyte molecule, but also on steric factors, and structural isomers are often resolved from one another.

Use of more polar solvents in the mobile phase will decrease the retention time of the analytes while more hydrophobic solvents tend to increase retention times. Particularly polar solvents in a mixture tend to deactivate the column by occupying the stationary phase surface. This is somewhat particular to normal phase because it is most purely an adsorptive mechanism (the interactions are with a hard surface rather than a soft layer on a surface).

NP-HPLC had fallen out of favour in the 1970s with the development of reversed-phase HPLC because of a lack of reproducibility of retention times as water or protic organic solvents changed the hydration state of the silica or alumina chromatographic media. Recently it has become useful again with the development of HILIC bonded phases which utilize a partition mechanism which provides reproducibility.

Reversed Phase Chromatography

Reversed phase HPLC (RP-HPLC) consists of a non-polar stationary phase and a moderately polar mobile phase. One common stationary phase is a silica which has been treated with RMe2SiCl, where R is a straight chain alkyl group such as $C_{18}H_{37}$ or C_8H_{17}. The retention time is, therefore, longer for molecules which are more non-polar in nature, allowing polar molecules to elute more readily.

Retention time is increased by the addition of polar solvent to the mobile phase and decreased by the addition of more hydrophobic solvent. Reversed phase chromatography is so commonly used that it is not uncommon for it to be incorrectly referred to as "HPLC" without further specification.

RP-HPLC operates on the principle of hydrophobic interactions which result from repulsive forces between a relatively polar solvent, the relatively non-polar analyte, and the non-polar stationary phase. The driving force in the binding of the analyte to the stationary phase is the decrease in the area of the non-polar segment of the analyte molecule exposed to the solvent.

This hydrophobic effect is dominated by the decrease in free energy from entropy associated with the minimization of the ordered molecule-polar solvent interface. The hydrophobic effect is decreased by adding more non-polar solvent into the mobile phase. This shifts the partition coefficient such that the analyte spends some portion of time moving down the column in the mobile phase, eventually eluting from the column.

The characteristics of the analyte molecule play an important role in its retention characteristics. In general, an analyte with a longer alkyl chain length results in a longer retention time because it increases the molecule's hydrophobicity.

Very large molecules, however, can result in incomplete interaction between the large analyte surface and the alkyl chain. Retention time increases with hydrophobic surface area which is roughly inversely proportional to solute size. Branched chain compounds elute more rapidly than their corresponding isomers because the overall surface area is decreased.

Apart from mobile phase hydrophobicity, other mobile phase modifiers can affect analyte retention. For example, the addition of inorganic salts causes a linear increase in the surface tension of aqueous solutions, and because the entropy of the analyte-solvent interface is controlled by surface tension, the addition of salts tends to increase the retention time.

Another important component is pH since this can change the hydrophobicity of the analyte. For this reason most methods use a buffering agent, such as sodium phosphate, to control the pH. An organic acid such as formic acid or most commonly trifluoro-acetic acid is often added to the mobile phase.

These serve multiple purposes: They control pH, neutralize the charge on any residual exposed silica on the stationary phase and act as ion pairing agents to neutralize charge on the analyte. The effect varies depending on use but generally improves the chromatography.

Reversed phase columns are quite difficult to damage compared with normal silica columns; however, many reverse phase columns consist of alkyl derivatized silica particles and should never be used with aqueous bases as these will destroy the underlying silica backbone. They can be used with aqueous acid but the column should not be exposed to the acid for too long, as it can corrode the metal parts of the HPLC equipment.

The metal content of HPLC columns must be kept low if the best possible ability to separate substances is to be retained. A good test for the metal content of a column is to inject a sample which is a mixture of 2, 2'- and 4, 4'- bipyridine. Because the 2, 2'- bipyridine can chelate the metal it is normal that when a metal ion is present on the surface of the silica the shape of the peak for the 2, 2'-bipyridine will be distorted, tailing will be seen on this distorted peak.

Size Exclusion Chromatography

Size exclusion chromatography (SEC) is a chromatographic method in which particles are separated based on their size, or in more technical terms, their hydrodynamic volume. It is usually applied to large molecules or macromolecular complexes such as proteins and industrial polymers. When an aqueous solution is used to transport the sample through the column, the technique is known as gel filtration chromatography.

The name gel permeation chromatography is used when an organic solvent is used as a mobile phase. The main application of gel filtration chromatography is the fractionation of proteins and other water-soluble polymers, while gel permeation chromatography is used to analyze the molecular weight distribution of organic-soluble polymers. Either technique should not be confused with gel electrophoresis, where an electric field is used to "pull" or "push" molecules through the gel depending on their electrical charges.

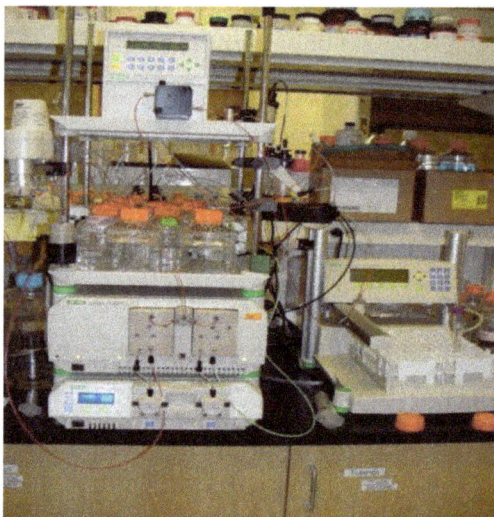

Equipment for running size exclusion chromatography. The buffer is pumped through the column (right) by a computer controlled device.

SEC is a widely used technique for the purification and analysis of synthetic and biological polymers, such as proteins, polysaccharides and nucleic acids. Biologists and biochemists typically use a gel medium—usually polyacrylamide, dextran or agarose—and filter under low pressure. Polymer chemists typically use either a silica or cross-linked polystyrene medium under a higher pressure. These media are known as the stationary phase.

The advantage of this method is that the various solutions can be applied without interfering with the filtration process, while preserving the biological activity of the particles to be separated. The technique is generally combined with others that further separate molecules by other characteristics, such as acidity, basicity, charge, and affinity for certain compounds.

The underlying principle of SEC is that particles of different sizes will elute (filter) through a stationary phase at different rates. This result in the separation of a solution of particles based on size, provided that all the particles are loaded simultaneously or near simultaneously, particles of the same size should elute together.

This is usually achieved with an apparatus called a column, which consists of a hollow tube tightly packed with extremely small porous polymer beads designed to have pores of different sizes. These pores may be depressions on the surface or channels through the bead. As the solution travels down the column some particles enter into the pores. Larger particles cannot enter into as many pores. The larger the particles, the less overall volume to traverse over the length of the column, and the faster the elution.

The filtered solution that is collected at the end is known as the eluent. The void volume consists of any particles too large to enter the medium, and the solvent volume is known as the column volume. In real life situations, particles in solution do not have a constant, fixed size, resulting in the probability that a particle which would otherwise be hampered by a pore may pass right by it. Also, the stationary phase particles are not ideally defined; both particles and pores may vary in size.

Elution curves, therefore, resemble Gaussian distributions. The stationary phase may also interact in undesirable ways with a particle and influence retention times, though great care is taken by column manufacturers to use stationary phases which are inert and minimize this issue.

Like other forms of chromatography, increasing the column length will tighten the resolution, and increasing the column diameter increases the capacity of the column. Proper column packing is important to maximize resolution: an over packed column can collapse the pores in the beads, resulting in a loss of resolution. An under packed column can reduce the relative surface area of the stationary phase accessible to smaller species, resulting in those species spending less time trapped in pores.

In simple manual columns the eluent is collected in constant volumes, known as fractions. The more similar the particles are in size, the more likely they will be in the same fraction and not detected separately. More advanced columns overcome this problem by constantly monitoring the eluent.

The collected fractions are often examined by spectroscopic techniques to determine the concentration of the particles eluted. Three common spectroscopy detection techniques are Refractive Index (RI), Evaporative Light Scattering (ELS), and Ultraviolet (UV). When eluting spectroscopically similar species (such as during biological purification) other techniques may be necessary to identify the contents of each fraction.

The elution volume decreases roughly linearly with the logarithm of the molecular hydrodynamic volume (often assumed to be proportional to molecular weight). Columns are often calibrated using 4-5 standard samples (e.g., folded proteins of known molecular weight) to determine the void volume and the slope of the logarithmic dependence. This calibration may need to be repeated under different solution conditions.

Applications

Proteomics

SEC is generally considered a low resolution chromatography as it does not discern similar species very well, and is, therefore, often reserved for the final "polishing" step of a purification. The technique can determine the quaternary structure of purified proteins which have slow exchange times, since it can be carried out under native solution conditions, preserving macromolecular interactions.

SEC can also assay protein tertiary structure as it measures the hydrodynamic volume (not molecular weight), allowing folded and unfolded versions of the same protein to be distinguished. For example, the apparent hydrodynamic radius of a typical protein domain might be 14a and 36A for the folded and unfolded forms respectively.

SEC allows the separation of these two forms as the folded form will elute much later due to its smaller size. Alternatively, folded and unfolded versions of the same metalloproteinase can be separated according to their different isoelectric points by using Quantitative Preparative Native Continuous Polyacrylamide Gel Electrophoresis (QPNC-PAGE).

Polymer Synthesis

SEC can be used as a measure of both the size and the polydispersity of a synthesised polymer; that one is able to find distribution of sizes of polymer molecules. If standards of a known size are run previously, then a calibration curve can be created to determine the sizes of polymer molecules of interest.

Alternatively, techniques such as light scattering and/or viscometry can be used online with SEC to yield absolute molecular weights that do not rely on calibration with standards of known molecular weight. Due to the difference in size of two polymers with identical molecular weights, the absolute determination methods are generally more desirable. A typical SEC system can quickly (in about half an hour) give polymer chemists information on the size and polydispersity of the sample.

Bio-affinity Chromatography

This chromatographic process relies on the property of biologically active substances to form stable, specific, and reversible complexes. The formation of these complexes involves the participation of common molecular forces such as the van der Waals interaction, electrostatic interaction, dipole- dipole interaction, hydrophobic interaction, and the hydrogen bond. An efficient, bio-specific bond is formed by a simultaneous and concerted action of several of these forces in the complementary binding sites.

Parameters

There are different parameters upon which the separation by HPLC relies. Some of the important parameters are discussed below.

Internal Diameter

The internal diameter (ID) of an HPLC column is a critical aspect that determines quantity of analyte that can be loaded onto the column and also influences sensitivity. Larger columns are usually seen in industrial applications such as the purification of a drug product for later use. Low ID columns have improved sensitivity and lower solvent consumption at the expense of loading capacity.

- Larger ID columns (over 10 mm) are used to purify usable amounts of material because of their large loading capacity.

- Analytical scale columns (4.6 mm) have been the most common type of columns, though smaller columns are rapidly gaining popularity. They are used in traditional quantitative analysis of samples and often use a UV-Vis absorbance detector.

In Gas-Phase Chromatography, two-dimensional separation is achieved by coupling a second, short column to the first long column. Coupling is achieved by different techniques, for example, shock-freezing the elutes in order of elution from the first column at fixed time-intervals, and then reheating them in order of elution, releasing them into the second column. The time of traversal through the second column needs to be shorter than the time remaining until the next sample is reheated to prevent compound build-up and to fully exploit the separational capability.

Pyrolysis Gas Chromatography

Pyrolysis-chromatography is a potent analytical tool able to thermally crack (fragment) essentially non-volatile molecules into fragments suitable for chromatographic analysis. The technique enables a reproducible and characteristic "fingerprint" to be generated of a non-volatile sample. The technique can be applied to such varied tasks as bacterial strain differentiation and forensic characterisation of paints, polymers and fibre cross-matching.

Type of Chromatography	Applications in the Real World	Why and What is it
Liquid Chromatography	Testing water samples to look for pollution.	Used to analyse metal ions and organic compound in solutions. It used liquids which may incorporated hydrophilic, insoluble molecules.
Gas Chromatography	Detecting bombs in airports, identifying and quantifying such drugs as alcohol, being used in forensics to compare fibres found on a victim.	Used to analyze volatile gases. Helium is used to move the gaseous mixture through a column of absorbent material.
Thin-layer Chromatography	Detecting pesticide or insecticide residues in food, also used in forensics to analyse the dye composition of fibres.	Uses an absorbent material on flat glass plates. This is a simple and rapid method to check the purity of the organic compound.
Paper Chromatography	Separating amino acids and anions, RNA fingerprinting, separating and testing histamines, antibiotics.	The most common type of chromatography The paper is the stationary phase. This uses capillary action to pull the solute up through the paper and separate the solute.

Adsorption Chromatography

Adsorption chromatography is probably one of the oldest types of chromatography around. It utilizes a mobile liquid or gaseous phase that is adsorbed onto the surface of a stationary solid phase. The equilibration between the mobile and stationary phases accounts for the separation of different solutes.

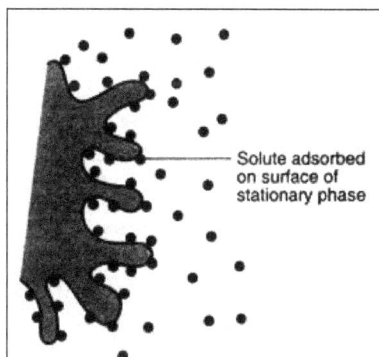

Adsorption Chromatography.

Partition Chromatography

This form of chromatography is based on a thin film formed on the surface of a solid support by a liquid stationary phase. Solute equilibrates between the mobile phase and the stationary liquid.

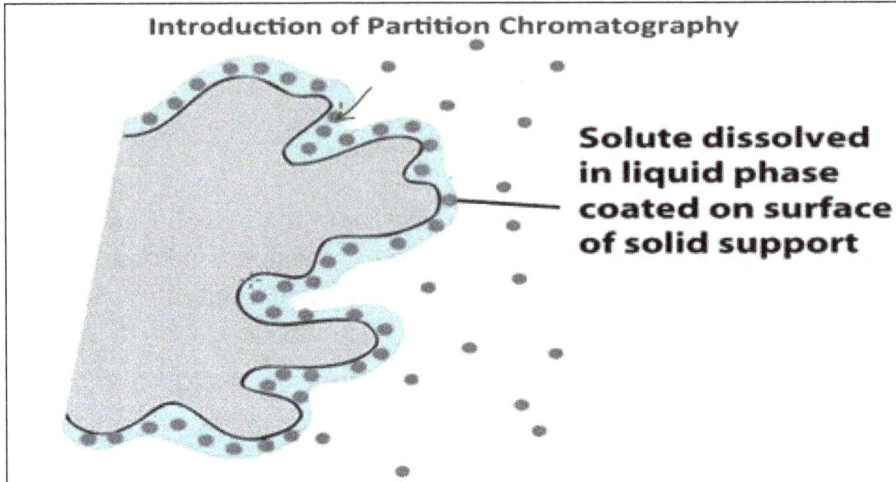

Introduction of Partition Chromatography

Solute dissolved in liquid phase coated on surface of solid support

Partition Chromatography.

Ion Exchange Chromatography

In this type of chromatography, the use of a resin (the stationary solid phase) is used to covalently attach anions or cations onto it. Solute ions of the opposite charge in the mobile liquid phase are attracted to the resin by electrostatic forces.

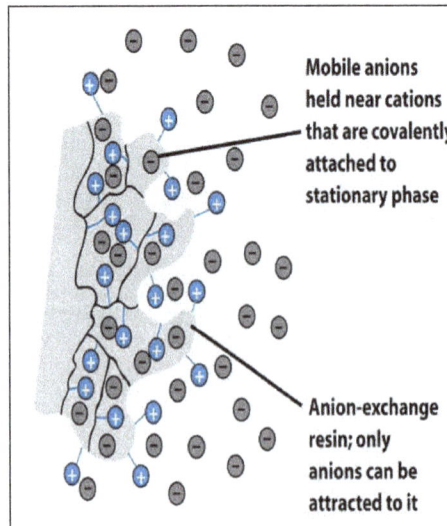

Mobile anions held near cations that are covalently attached to stationary phase

Anion-exchange resin; only anions can be attracted to it

Molecular Exclusion Chromatography

Also known as gel permeation or gel filtration, this type of chromatography lacks an attractive interaction between the stationary phase and solute. The liquid or gaseous phase passes through a porous gel which separates the molecules according to its size. The pores are normally small and exclude the larger solute molecules, but allow smaller molecules to enter the gel, causing them to flow through a larger volume. This causes the larger molecules to pass through the column at a faster rate than the smaller ones.

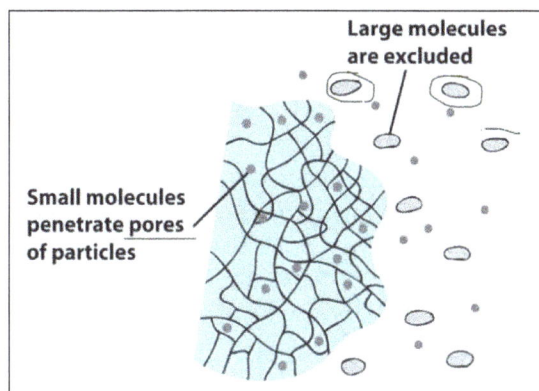

Molecular Exclusion Chromatography.

Affinity Chromatography

This is the most selective type of chromatography employed. It utilizes the specific interaction between one kind of solute molecule and a second molecule that is immobilized on a stationary phase. For example, the immobilized molecule may be an antibody to some specific protein. When solute containing a mixture of proteins are passed by this molecule, only the specific protein is reacted to this antibody, binding it to the stationary phase. This protein is later extracted by changing the ionic strength or pH.

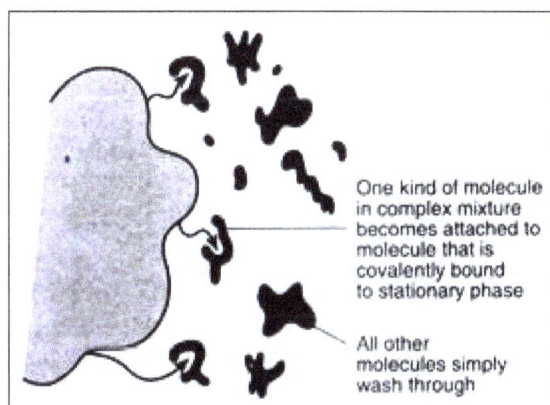

Affinity Chromatography.

Counter-Current Chromatography

CCC or Partition Chromatography is a category of liquid-liquid chromatography techniques. Chromatography in general is used to separate components of a mixture based on their differing affinities for mobile and stationary phases of a column. The components can then be analyzed separately by various sorts of detectors which may or may not be integrated into an apparatus. In liquid-liquid chromatography, both the mobile and stationary phases are liquid.

In contrast, standard column chromatography uses a solid stationary phase and a liquid mobile phase, while gas chromatography uses a liquid stationary phase on a solid support and a gaseous mobile phase. By eliminating solid supports, permanent adsorption of the analyte onto the column is avoided, and a near 100% recovery of the analyte can be achieved.

The instrument is also easily switched between various modes of operation simply by changing solvents. With liquid-liquid chromatography, researchers are not limited by the composition of the columns commercially available for their instrument. Nearly any pair of immiscible solutions can be used in liquid-liquid chromatography, and most instruments can be operated in standard or reverse-phase modes. Solvent costs are also generally cheaper than that HPLC, and the cost of purchasing and disposing of solid adsorbents is completely eliminated.

Another advantage is that experiments conducted in the lab can easily be scaled to industrial volumes. When GC or HPLC is done with large volumes, resolution is lost due to issues with surface-to-volume ratios and flow dynamics; this is avoided when both phases are liquid.

CCC can be thought of as occurring in three stages: mixing, settling, and separation (although they often occur continuously). Mixing of the phases is necessary so that the interface between them has a large area, and the analyte can move between the phases according to its partition coefficient.

A partition coefficient is a ratio of the amount of analyte found in each of the solvents at equilibrium and is related to the analyses' affinity for one over the other. The mobile phase is mixing with them settling from the stationary phase throughout the column.

The degree of stationary phase retention (inversely proportional to the amount of stationary phase loss or "bleed" in the course of a separation) is a crucial parameter. Higher quality instruments have greater stationary phase retention. The settling time is a property of the solvent system and the sample matrix, both of which greatly influence stationary phase retention.

There are certain modifications of CCC. Major of them are:

1. Droplet Counter-current Chromatography (DCCC):

Droplet CCC is the oldest form of CCC. It uses only gravity to move the mobile phase through the stationary phase. In descending mode, droplets of the more dense mobile phase and sample are allowed to fall through a column of the lighter stationary phase using only gravity. If a less dense mobile phase is used it will rise through the stationary phase, this is called ascending mode.

The eluent from one column is transferred to another; the more columns that are used, the more theoretical plates can be achieved. The disadvantage of DCCC is that flow rates are low, and poor mixing is achieved for most binary solvent systems, which make this technique both time-consuming and inefficient.

2. Centrifugal Partition Chromatography (CPC):

CPC uses centrifugal force to speed separation and achieve greater flow rates than DCCC (which relies on gravity) (e.g., the Kromaton FCPC). The columns are cut into a rotor, oriented out from the middle, and connected by channels. The rotor is filled with the stationary phase, and the mobile phase is pumped through it as the rotor spins.

Early CPC instruments had the disadvantage of requiring a rotary seal that frequently needed to be replaced, but a new generation of instruments is now reliable. CPC can also be operated in either descending or ascending mode, where the direction is relative to the force generated by the rotor rather than gravity.

3. High-Speed Counter-current Chromatography:

The modern era of CCC began with the development by Dr. Yoichiro Ito of the planetary centrifuge and the many possible column geometries it can support. These clever devices make use of a little- known means of making non-rotating connections between the stator and the rotor of a centrifuge.

Functionally, the high-speed CCC consists of a helical coil of inert tubing which rotates on its planetary axis and simultaneously rotates eccentrically about another solar axis. The effect is to create zones of mixing and zones of settling which progress along the helical coil at dizzying speed. This produces a highly favorable environment for chromatography.

There are numerous potential variants upon this instrument design. The most significant of these is the toroidal CCC (e.g., the Pharma-Tech TCC-1000). This instrument does not employ planetary motion. In some respects it is very alike CPC, but retains the advantage of not needing rotary seals.

It also employs a capillary tube instead of the larger-diameter tubes employed in the helices of the other CCC models. This capillary passage makes the mixing of two phases very thorough, despite the lack of shaking or other mixing forces. This instrument provides rapid analytical scale separations, which can nonetheless be scaled up to either of the larger-scale CCC instruments.

Modes of Operation

CCC can be operated in different modes depending on the type of separation required and sample to be separated.

These different operation modes of CCC are as under:

1. Head to tail: The denser phase is pumped through as the mobile phase. Derived from terminology for Archimedean screw force.

2. Tail to head: The less dense phase is used as the mobile phase.

3. Dual Mode: The mobile and stationary phases are reversed part way through the run.

4. Gradient Mode: The concentration of one or more components in the mobile phase is varied throughout the run to achieve optimal resolution across a wider range of polarities. For example, a methanol-water gradient may be employed using pure heptane as the stationary phase. This is not possible with all binary systems due to excessive loss of stationary phase.

5. Elution Extrusion Mode (EECCC): The mobile phase is extruded after a certain point by switching the phase being pumped into the system. For example, during the elution portion of a separation using an EtOAc-water system running head to tail, the aqueous mobile phase is being pumped into the system. In order to switch to extrusion mode, organic phase is pumped into the system.

    This can be accomplished either with a valve on the inlet of single pump, or ideally with an integrated system of two or three pumps, each dedicated either to a single phase of a binary mixture, or to an intermediate wash solvent. This also allows for good resolution of compounds with high mobile-phase affinities. It requires only one column volume of solvent and leaves the column full of fresh stationary phase.

6. pH Zone Refining: Acidic and basic solvents are used to elute analytes based on their pKa.

Principles of Chromatography

The molecules present in biological system or in synthetic chemistry are produced through a series of reactions involving intermediates. At any moment of time biological organism has major fraction as desired product but has other compounds in minute quantities. The minor species present in a product is always referred as "impurities" and these compounds need to separate from desired product for biotechnology applications. How two molecules can be separated from each other? To answer this question we can take the example of three molecules given in figure. These 3 molecules (benzene, phenol, aniline) are similar to each other but have distinct physical and chemical properties which can be used as a criteria to separate them. The physical and chemical properties which can be use to separate molecules are:

Physical Properties

1. Molecular weight;
2. Boiling point (in case both are liquid, as in this case);
3. Freezing point;
4. Crystallization;
5. Solubility;
6. Density.

Chemical Properties

1. Functional Group, for example, phenol has –OH where as aniline has NH_2;
2. Reactivity towards other reagent to form complex.

Now for example you have a mixture of compound 1 (benzene) and compound 3 (Aniline) and you would like to purify benzene rather than aniline. In this situation, you can take the physical and chemical properties of benzene into the account and isolate it from the mixture.

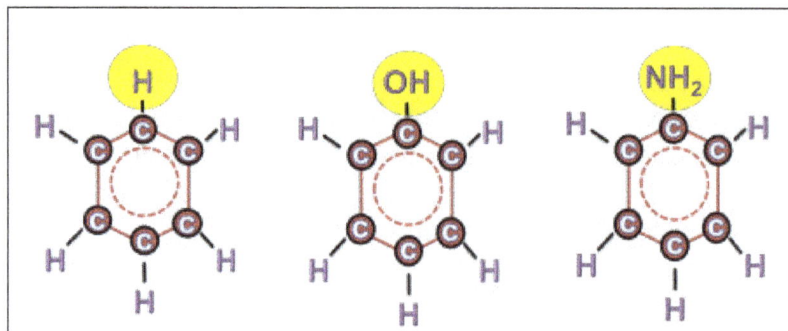

Table: Chemical Structure and physical Properties of benzene, phenol and aniline.

Name	Benzene	Phenol	Aniline
Molecular Formula	C_6H_6	C_6H_6O	$C_6H_5NH_2$
Molar mass (g mol^{-1})	78.11	94.11	93.13
Density	0.8765 g cm^{-3}	1.07 g cm^{-3}	1.0217 g ml^{-1}
Melting point (°C)	5.5	4.5	-6.3
Boiling point (°C)	80.1	181.7	184.13

Principle of Seperation

How a physical or chemical property will allow to isolate a particular substance? The mixture of compound 1 and 3 is shown in figure and assume if we are using boiling point as a criteria to isolate them. As we will heat the mixture there will two phase forms, one liquid phase and other is vapor phase. The molecules of compound 1 and 3 will distribute between these two phases and as the temp is near to boiling point of compound 1, more amount of 1 will be present in vapor phase than liquid phase. Where as more number of compound 3 will be in liquid phase. Eventually as this process will continue, at the end two molecules will get separated from each other. The distribution coefficient (Kd) to describe the distribution of compound 1 between two phase A and B is as follows:

$$Kd = \frac{\text{Concentration in Phase A}}{\text{Concentration in Phase B}}$$

Similarly one can also exploit other physical & chemical parameters as well. With each and every physical and chemical parameter the molecule present in the mixture will distribute as per their behavior in each parameter.

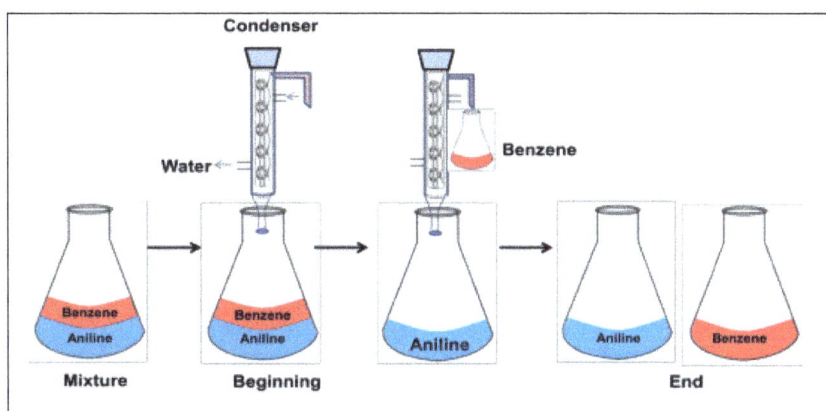

Distribution of molecules during distillation.

The purpose of chromatography is to separate a complex mixture into individual component exploiting the partition effect which distribute the molecules into the different phases. As discussed, a distribution of a molecule between two phases A and B is given by a distribution coefficient, Kd. In most of the chromatography techniques, phase A is stationary phase or matrix and phase B is mobile phase or buffer.

Column Chromatography

In column chromatography, a stationary phase is filled into a cylindrical tube made up of glasss or steel. The mixture of analyte is loaded on the top and it runs from top to bottom. How Kd is exploited in column chromatography? Assume two molecules, X and Y with a Kd value of 1 and 9 and they are traveling through a column with water as mobile phase as given in figure. As they will travel, X and Y will partition between stationary phase and mobile phase. As there is a huge difference in Kd, Y will be associated with the matrix and remain on the top of the column where as X will move along the water. At the end of chromatography, X will come out first whereas Y will come out last.

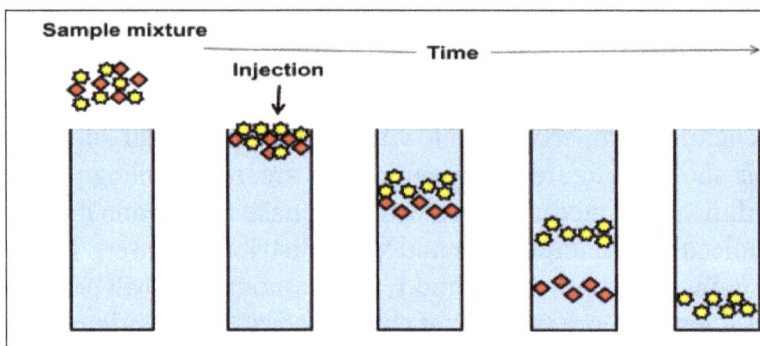

Seperation of two molecules on a column.

Chromatogram

The plot of elution volume along with the absorbance is known as chromatogram as given in figure. The volume or time it takes for a analyte to come out from the column is known as retention volume or time. The chromatogram may have separate peaks (A and B) or peaks (C and D) with overlapping base, these peaks are called fused peaks.

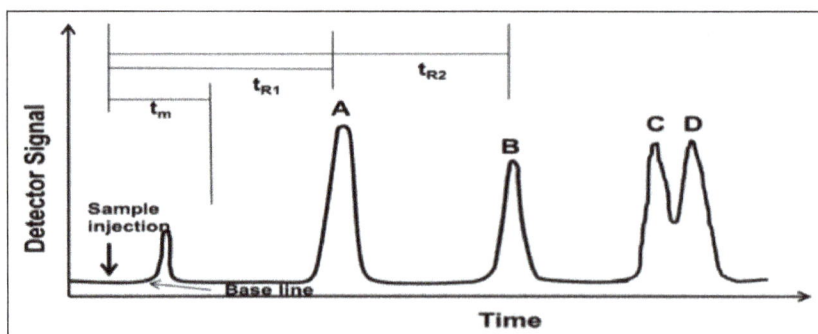

A typical chromatogram.

Resolution: The ability of a chromatography column to separate two analyte peak from another is known as resolution. It is defined as the ratio of difference in retention time between two peaks and average of base of peak width. It is given by,

$$Rs = \frac{\Delta tR}{Wav}$$

When Rs=1, the separation of two peaks is 97.7% and a column with Rs more than 1.5 considered good. The number of distribution events govern the ability of a column to separate the two analytes. In another words, resolution is directly proportional to the number of distribution events. In column chromatography, each thin plain of column matrix participate in distribution of molecule. Assume height of a distribution plain is H and length of a column is L, hence number (N) of distribution plain in a column is given by,

$$N = \frac{L}{H}$$

$$N = 16\left(t_R / W\right)^2$$

$$N = 5.54\left(t_R / W_{1/2}\right)^2$$

Hence, Number of distribution plain in a column is controlling two parameters:

- As number of distribution plain will go up, it will allow the analyte to travel for longer period of time, consequently it will increase the distance between two peaks.

- As number of distribution plain will go up, it will reduce the width of the base of peak, as a result the peaks will be more sharp. A representative example, how number of distribution plain affects the base of the peak is given in figure. As the number is increasing, the peak width is decreasing. Hence, number of distribution is an indirect way to measure the column efficiency, higher N number is desirable for better separation.

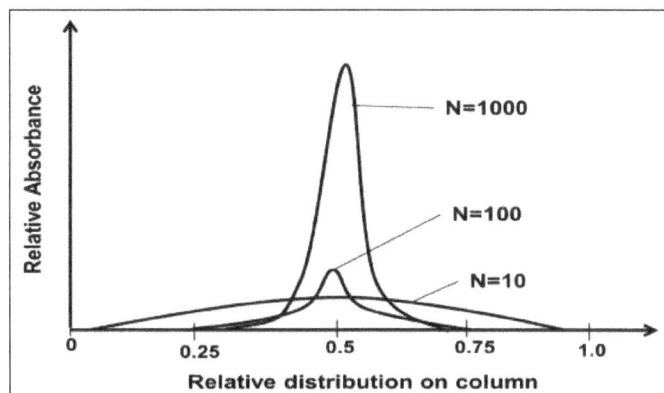

Relationship between number of distribution planes (N) and peak width.

Different components of chromatography system: The different components of a chromatography system are given in figure. It has following components:

- Reservoir: One or two reservoir for mobile phase (buffer).

- Pump: One or two pump to flow the buffer from reservoir. Different types of pumps are used in chromatography system, mostly based on the pressure level required to perform chromatography. A pump is chosen as per the pressure required to run the mobile phase. Based on the pressure level, liquid chromatography can be classified into the following categories:

  ○ Low Pressure Liquid Chromatography: Pressure limit less than 5 Bar.

  ○ Medium Pressure Liquid Chromatography: Intermediate pressure limit (6-50 bar).

  ○ High Pressure Liquid Chromatography: Pressure limit more than 50-350 bar.

A typical polysaccharide bead is not appropriate to withstand high pressure during HPLC. Hence, in HPLC silica based beads are recommended. Due to high pressure and smaller size of the silica beads gives higher number of theoretical plates. This gives HPLC superior resolving power to separate complex biological samples.

1. Mixer: A mixer is required to mix the buffer received from both pumps to form a linear or step gradient.

2. Column: A column made up of glass or steel.

3. Detector: The elution coming out from column goes to the online monitoring system to test the presence of the analyte based on different properties. There are different types of detectors are known in chromatography such as UV-Visible detector etc.

4. Fraction Collection: The eluent can be collected in different fractions by a fraction collector.

5. Recorder: The profile of eluent with respect to the measured property in a detector can be plotted in the recorder.

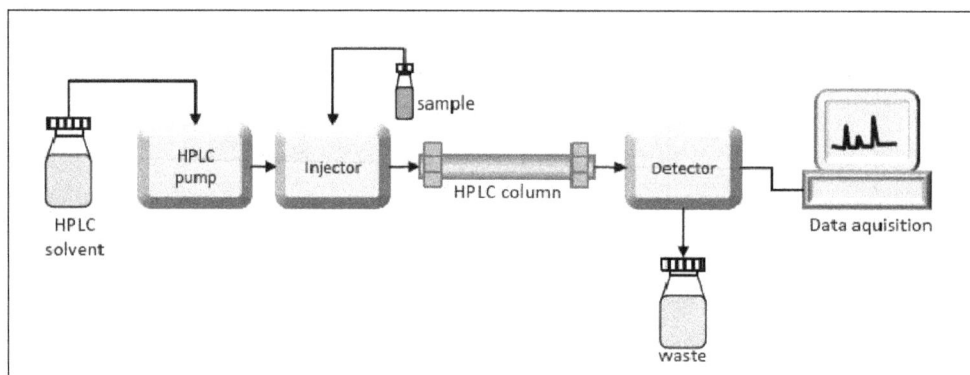

Different components of a chromatography system.

Different Forms of Chromatography

Partition Chromatography: In this form of chromatography, an analyte distribute themselves into two phases, liquid stationary and mobile phase. The major advantage of this chromatography is that it is simple, low cost and has broad specificity. It is further divided into liquid-liquid chromatography and bonded-phase liquid chromatography. The example of this chromatography is cellulose, starch or silica matrix.

Adsorption Chromatography: In this form of chromatography, matrix molecule has ability to hold the analyte on their surface through a mutual interaction due to different types of forces such as hydrogen bonding, electrostatic interaction, vander waal etc. The example are ion-exchange chromatography, hydrophobic interaction chromatography, affinity chromatography etc.

Ion-Exchange Chromatography-I

Ion exchange chromatography: Ion-exchange chromatography is a versatile, high resolution chromatography techniques to purify the protein from a complex mixture. In addition, this chromatography has a high loading capacity to handle large sample volume and the chromatography operation is very simple.

Principle: This chromatography distributes the analyte molecule as per charge and their affinity towards the oppositely charged matrix. The analytes bound to the matrix are exchanged with a competitive counter ion to elute. The interaction between matrix and analyte is determined by net charge, ionic strength and pH of the buffer. For example, when a mixture of positively charged analyte (M, M^+, M^{-1}, M^{-2}) loaded onto a positively charged matrix, the neutral or positively charged analyte will not bind to the matrix where as negatively charged analyte will bind as per their relative charge and needed higher concentration of counter ion to elute from matrix.

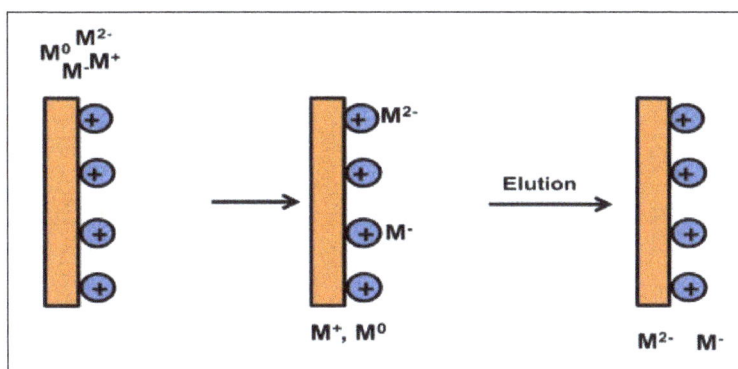

Affinity of analytes (M, M^+, M^{-1}, M^{-2}) towards positively charged matrix.

The matrix used in ion-exchange chromatography is present in the ionized form with reversibly bound ion to the matrix. The ion present on matrix participitate in the reversible exchange process with analyte. Hence, there are two types of ion-exchange chromatography:

1. Cation exchange chromatography- In cation exchange chromatography, matrix has a negatively charged functional group with a affinity towards positively charged molecules. The positively charged analyte replaces the reversible bound cation and binds to the matrix. In the presence of a strong cation (such as Na^+) in the mobile phase, the matrix bound positively charged analyte is replaced with the elution of analyte. The popular cation exchangers used are given in table.

2. Anion Exchange chromatography- In anion exchange chromatography, matrix has a positively charged functional group with a affinity towards negatively charged molecules. The negatively charged analyte replaces the reversible bound anion and binds to the matrix. In the presence of a strong anion (such as Cl^-) in the mobile phase, the matrix bound negatively charged analyte is replaced with the elution of analyte. The popular anion exchangers used are given in table.

List of selected Ion-exchange matrix			
S.No	Name	Functional Group	Type of Ion-exchanger
1	Carboxyl methyl (CM)	$-OCH_2COOH$	Cation Exchanger
2	Sulphopropyl (SP)	$-OCH_2CH_2CH_2SO_3H$	Cation Exchanger
3	Sulphonate (S)	$-OCH_2SO_3H$	Cation Exchanger
4	Diethylaminoethyl (DEAE)	$-OCH_2CH_2NH(C_2H_5)_2$	Anion Exchanger
5	Quaternary aminomethyl (Q)	$-OCH_2N(CH_3)_3$	Anion Exchanger

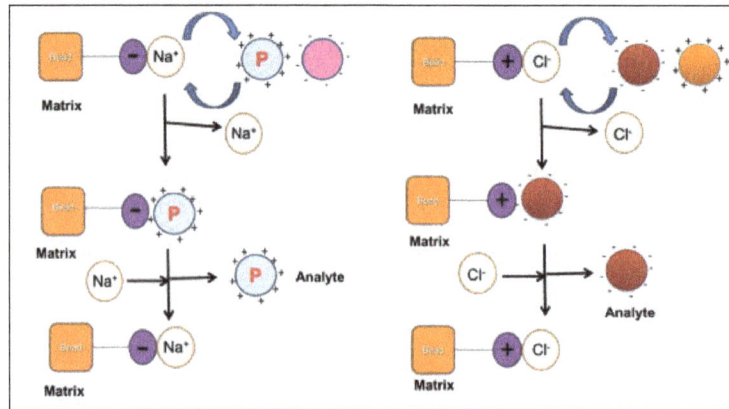

Cation and Anion exchange chromatography.

Isoelectric point and charge on a protein: Protein is a polymer made up of amino acids with ionizable side chain. At a particular pH, these amino acid side chain ionizes differently to give a net charge (positive/negative) to the protein. The pH at which the net charge on a protein is zero is called as Isoelectric point (pI). The protein will have a net positive charge below the pI where as it has net negative charge above the pI value.

Choice of a Ion-exchange column matrix-Before starting the isolation and purification of a substance, a choice for a suitable ion-exchange chromatography is important. There are multiple parameter which can be consider for choosing the right column matrix.

1. pI value and Net charge: The information of a pI will be allow you to calculate the net charge at a particular pH on a protein. As discussed above, a cation exchange chromatography can be use below the pI where as an anion exchange chromatography can be use above the pI value.

2. Structural stability: 3-D structure of a protein is maintained by electrostatic and vander waal interaction between charged amino acid, Π-Π interaction between hydrophobic side chain of amino acids. As a result, protein structure is stable in a narrow range around its pI and a large deviation from it may affect its 3-D structure.

3. Enzymatic activity: Similar to structural stability, enzymes are active in a narrow range of pH and this range should be consider for choosing an ion-exchange chromatography.

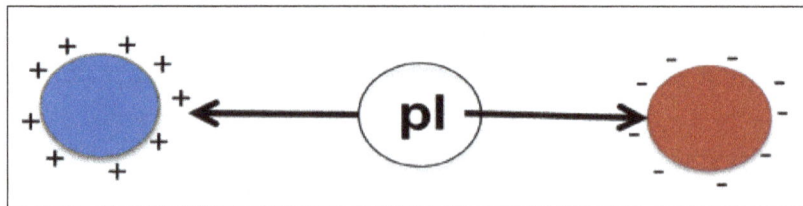

Change of charge with respect to the pI.

Operation of the technique-Several parameters needs to be consider to perform ionexchange chromatography.

1. Column material and stationary phase-Column material should be chemically inert to avoid destruction of biological sample. It should allow free low of liquid with minimum clogging. It should be capable to withstand the back pressure and it should not compress or expand during the operation.

2. Mobile Phase: The ionic strength and pH are the crucial parameters to influence the property of the mobile phase.

3. Sample Preparation: The sample is prepared in the mobile phase and it should be free of suspended particle to avoid clogging of the column. The most recommended method to apply the sample is to inject the sample with a syringe.

4. Elution: There are many ways to elute a analyte from the ion-exchange column. (1) Isocratic elution (2) Step-wise gradient (3) Continuous gradient either by salt or pH (4) affinity elution (5) displacement chromatography.

5. Column Regeneration: After the elution of analyte, ion-exchange chromatography column require a regeneration step to use next time. column is washed with a salt solution with a ionic strength of 2M to remove all non-specifically bound analytes and also to make all functional group in a iononized form to bind fresh analyte.

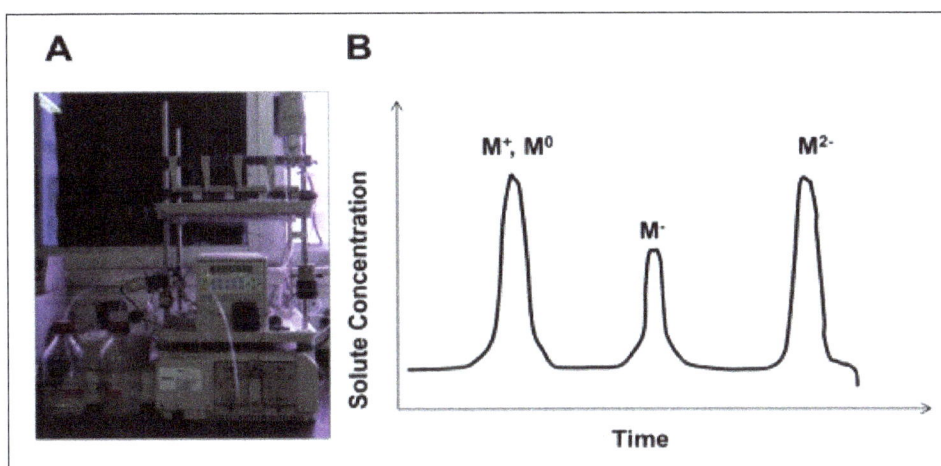

Operation of the Ion-exchange Chromatography. (A) Chromatography system to perform gradient elution of analytes to give an (B) elution profile.

Ion-Exchange Chromatography-II

Applications of Ion-exchange chromatography:

1. Protein Purification.

2. Protein-DNA interaction: Ion-exchange column is used as a tool to study interaction between DNA and a particular protein. DNA is negatively charged and has strong affinity towards anion exchange chromatography. A schematic figure to depict the steps involved in DNA-protein interaction is given in. In this approach, anion exchange matrix is incubated with the DNA and allowed it to bind tightly. Excesss DNA is washed from the column. Now the pure protein is passed through the DNA bound beads, followed by washing with the buffer to remove unbounded proteins. Now the DNA is eluted from the matrix either by adding high salt concentration or with denaturating condition. Now the fractions are tested for the presence of DNA and protein. Eluted protein is analyzed in the SDS-PAGE and DNA is in agarose. As a control, protein is also added to the matrix without DNA to rule out the possibility of protein binding directly to the matrix. If protein will have a affinity towards

DNA, they both will comes out from the column at the same time and should give similar pattern in the elution profile. It could be possible that high salt may break interaction between DNA and protein, in such situation protein will comes out first followed by DNA. Besides this ion-exchange chromatography approach still be able to answer whether the DNA-protein are interacting with each other or not.

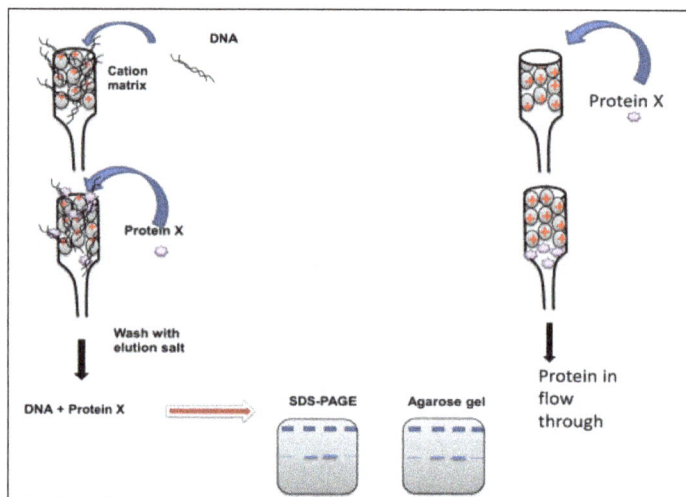

3. Softening of water: Ground water has several metals such as Ca^{2+}, Mg^{2+} and other cationic metals. Due to presence of the metal, hard water creates problem in industrial settings. Ion-exchange chromatography is used to remove the metals present in the water through an exchange of matrix bound Na^+. Calcium or magnesium present in the hard water has more affinity towards the matrix and it replaces with matrix bound sodium ions. The schematic presentation of water softening is given in figure. A cation exchanger matrix with bound sodium is packed in the column and the hard water containing calcium, magnesium is passed through the column. In this process, calcium present in the solution preferentially migrate from the solution to the matrix where as sodium ion present on the matrix migrate into the solution. The matrix can be use for softening of the water until it has bound sodium ions. Once sodium ions are exhausted, matrix can be regenerated by flowing a solution of sodium chloride or sodium hydroxide. The calcium/magnesium bound to the matrix comes out in the solution and can be dumped into the sewage.

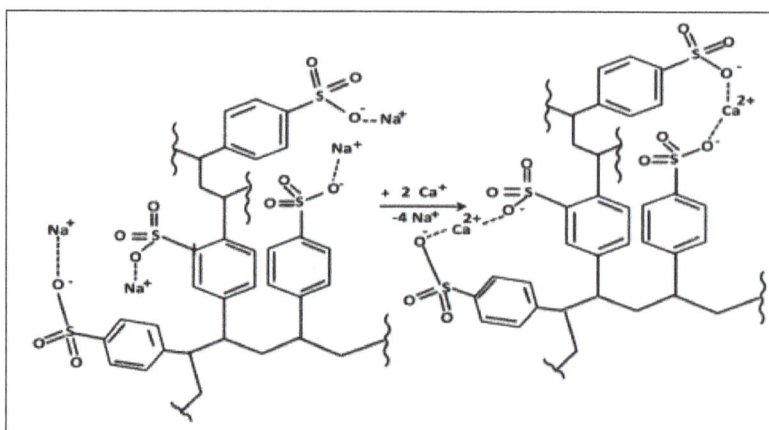

Mechanism of metal exchange during water softening.

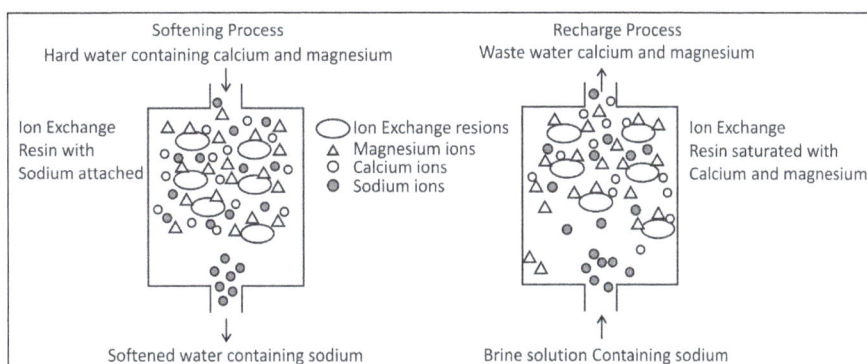

Softening of water by a cation exchanger matrix column.

4. Protein kinase assay: Protein kinase are class of enzyme responsible for transfer of phosphate group on the substrate molecule. In the protein kinase assay, a radioactive substrate (preferable a radioactivity on carbon) was incubated with the enzyme protein kinase, $MgCl_2$ and non-radioactive ATP. A negative control is also been included where enzyme protein kinase is absent from the assay mixture. Reaction mixture from negative control and experimental will be loaded on two separate cation exchange chromatography columns to bind unphosphorylated substrate from the reaction mixture where as phosphorylated radioactive substrate is present in the flow through. The radioactive count of the flow through was measured using a liquid scintillation reagent.

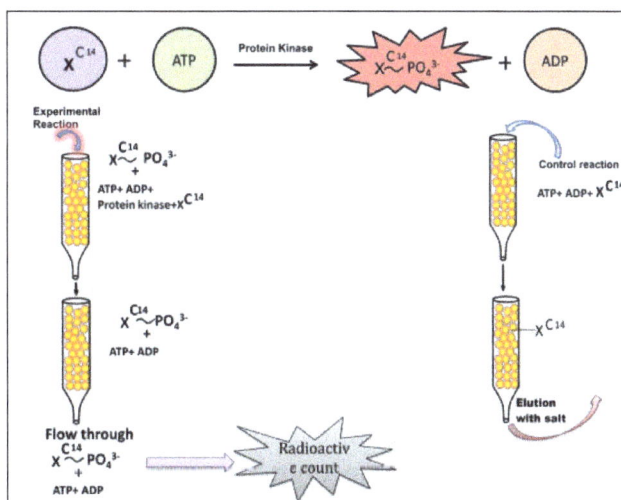

Protein kinase assay using ion-exchange chromatography.

5. Purification of rare earth metals from nuclear waste: Ion-exchange matrix is used to isolate and purify rare earth metals such as uranium or plutonium. The first process to isolate uranium in large quantities was developed by frank spedding. Ion- exchange beads are also found suitable to recover uranium from the water coming out of the nuclear power plant. Uranium binds to the matrix through the ion-exchange process. The uranium bound bead is sent to the processing unit where uranium is isolated from the beads to form 'yellow cake' and stored in drum for further processing. The ion-exchange beads can be reused in the ion-exchange facility.

6. Concentrating a sample: A ion-exchange bead can be used to bind the analyte from a diluted solution and then sample can be eluted in smaller volume to increase the concentration.

Hydrophobic Interaction Chromatography

Hydrophobic interaction chromatography exploits the ability of a strong interaction between hydrophobic group attached to the matrix and hydrophobic patches present on an analyte such as protein. Protein is made-up of amino acids with acidic, basic, polar and non-polar (aliphatic or aromatic) side chain. Protein is synthesized from ribosome as a linear chain and afterwards it gets folded into a 3-D conformation mostly guided by the environment of side chain and the outer medium. Local environment in a cell is aqeous and it favors the folding of protein to keep the polar or charged amino acids on the surface and non-polar side chain within the inner core. Most of the hydrophobic amino acids are shielded from the outer polar environment where as polar amino acid present on the surface has bound water molecule to form a hydration shell.

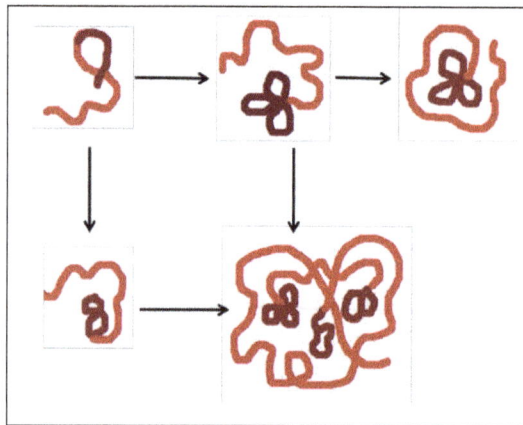

Folding of Protein in an aqueous environment. Following a series of folding stages, protein adopts a 3-D conformation with hydrophobic patches present in the core.

Addition of low amount of salt to the protein solution results in the displacement of bonded water molecule with an increase in protein solubility. This effect is called as "salting-in". In the presence of more amount of salt, water molecule shielding protein side chains are displaced completely with an exposure of hydrophobic patches on protein surface to induce protein precipitation or decrease in protein solubility. This effect is called as "salting-out". The phenomenon of salting out is modulated so that addition of salt induces exposure of hydrophobic patches on protein but does not cause precipitation or aggregation. The exposure of hydrophobic patches facilitates the binding of protein to the non-polar ligand attached to the matrix. When the concentration of salt is decreased, the exposed hydrophobic patches on protein reduces the affinity towards matrix and as a result it get eluted.

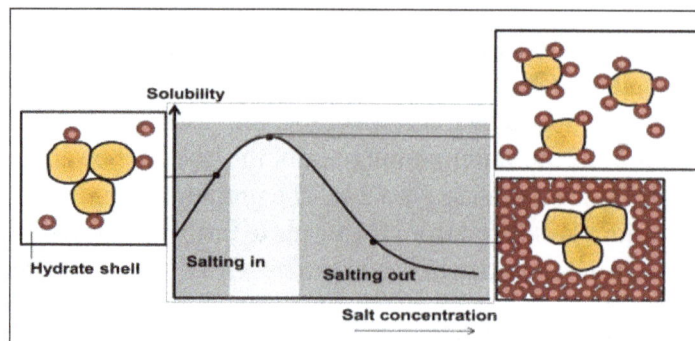

Effect of salt on protein, salting in and salting out effect.

The choice of HIC gel-The different commercially available HIC matrix are given in Table. Choosing a suitable HIC matrix is essential to achieve best result. The strength of the binding of analyte on a HIC column is governed by the length of the aliphatic linear ligand. Matrix with aromatic ring containing ligand makes additional Π-Π interaction and they will bind analyte more strongly than same number of carbon aliphatic ligand. In addition, presence of Π-Π interaction gives selectivity as well, such as ring containing aromatic ligand, phenylalanine. At last, ligand density plays a vital role in the strength of binding of an analyte to the matrix. Hence, these points should be consider to choose a suitable matrix for purification.

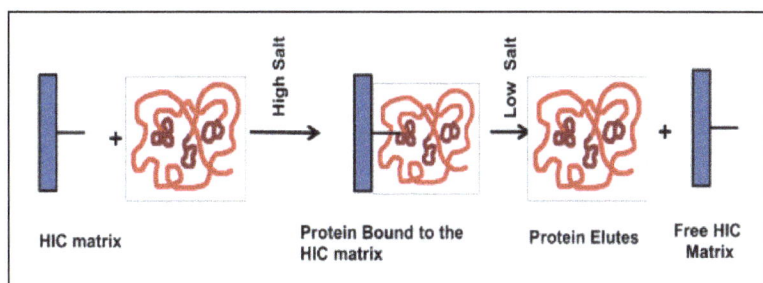

Principle of the hydrophobic interaction chromatography.

Selected list of popular HIC column matrix.		
S.No.	Column Material	Functional Group
1	Butyl-S-Sepharose	-Butyl
2	Phenyl Sepharose (Low Sub)	-Phenyl, low density
3	Phenyl Sepharose (High Sub)	-Phenyl, high density
4	Capto phenyl sepharose	-Phenyl
5	Octyl Sepharose	-Octyl

Operation of the technique-Several parameters needs to be consider to perform hydrophobic interaction chromatography.

1. Equlibration: HIC column material packed in a column and equilibrate with a buffer containing 0.5-1.5M ammonium sulphate (mobile phase). The salt must be below the concentration where it has salting-out effect.

2. Sample Preparation: The sample is prepared in the mobile phase and it should be free of suspended particle to avoid clogging of the column. The most recommended method to apply the sample is to inject the sample with a syringe.

3. Elution: There are many ways to elute a analyte from the hydrophobic interaction column. (1) decreasing salt concentration, (2) changing the polarity of the mobile phase such as alchol, (3) By a detergent to displace the bound protein.

4. Column Regeneration: After the elution of analyte, HIC column require a regeneration step to use next time. column is washed with 6M urea or guanidine hydrochloride to remove all non-specifically bound protein. The column is then equileberated with mobile phase to regenerate the column. The column can be stored at 4 °C in the presence of 20% alcohol containing 0.05% sodium azide.

Gel Filtration Chromatography

This chromatography distributes the protein or analyte, based on their size by passing through a porous beads. The first report in 1955 described performing a chromatography column with swollen gel of maize starch to separate the protein based on their size. 'Porath and Floidin'coined the term "gel filtration" for this chromatography technique separating the analytes based on molecular sizes. Since then the chromatography technique evolved in terms of developed of different sizes beads to separate protein of narrow range, as well as performing the technique in aqueous and non-aqueous mobile phase. The beads used in gel filtration chromatography is made up of cross linked material (such as dextran in sephadex) to form a 3-D mesh. These 3-D mesh swell in the mobile phase to develop pores of different sizes. The extent of cross linking controls the pores size within the gel beads.

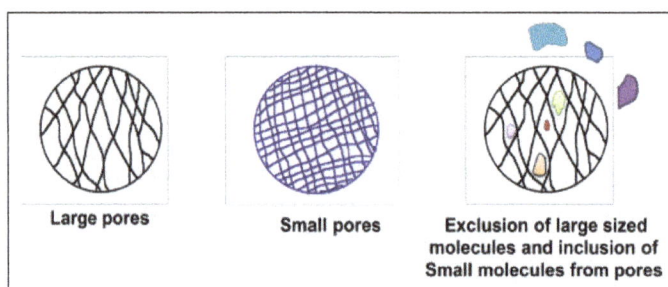

Gel Filtration Matrix has Beads with different pore sizes.

The principle of the chromatography technique is illustrated in figure. The column is packed with the beads containing pores to allow entry of molecules based on their sizes. Smallest size in the inner part of pore followed by gradual increasing size and largest molecule excluded from entering into the gel. The separation between molecules occur due to the time they travel to come out from the pores. When the mobile phase pass through the column, it takes protein along with it. The small molecules present in the inner part of the gel takes longer flow of liquid (or time) and travel longer path to come out where as larger molecules travel less distance to come out. As a result, the large molecule and small molecule get separated from each other. A schematic gel filtration chromatogram is given in figure.

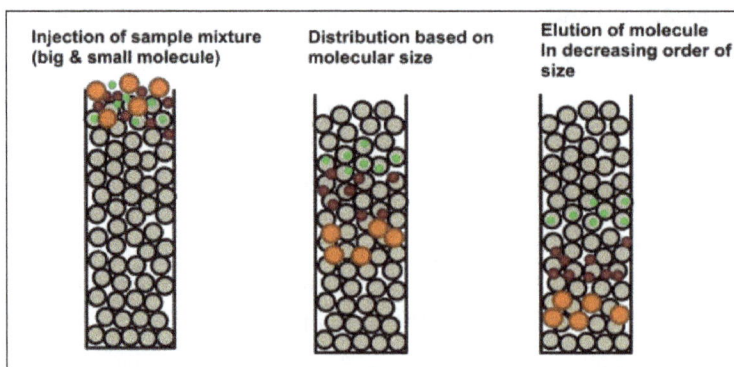

Principle of Gel Filtration Chromatography.

Suppose the total column volume of a gel is V_t and then it is given by:

$$V_t = V_g + V_i + V_o$$

V_g is the volume of gel matrix, V_i is the pore volume and V_o is the void volume. The volume of mobile phase flow to elute a column from a column is known as elution volume (V_e). The elution volum is related to the void volume and the distribution coefficient Kd as given below:

$$V_e = V_o + KdV_i$$

$$Kd = \frac{V_e - V_o}{V_i}$$

Kd is the ratio of inner volume available for an analyte and it is independent to the column geometry or length. As per relationship given in $Kd = \frac{V_e - V_o}{V_i}$, three different type of analytes are possible:

1.   Analyte with Kd=0, or $V_e=V_o$, these analytes will be completely excluded from the column.

2.   Analyte with Kd=1 or $V_e=V_o+V_i$, these analytes will be completely in the pore of the column.

3.   Analyte with Kd>1, in this situation analyte will adsorb to the column matrix.

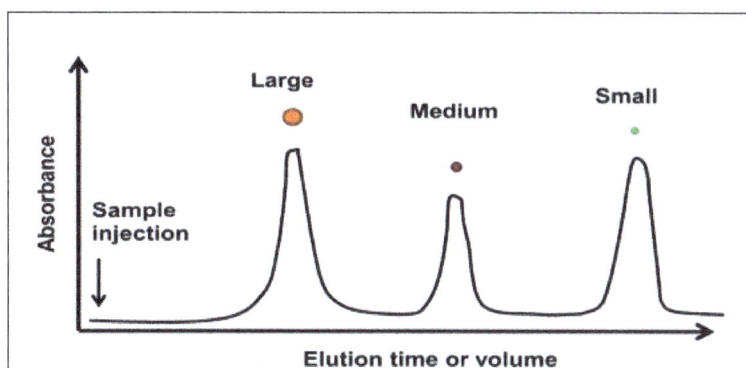

A typical Gel Filtration Chromatogram.

Choice of matrix for gel filtration chromatography-The choice of the column depends on the range of molecular weight and the pressure limit of the operating equipment. A list of popular gel filtration column matrix with the fractionation range are given in table.

Operation of the chromatography:

1.   Column packing: The column material is allowed to swell in the mobile phase. It is poured into the glass tube and allow the beads to settle without trapping air bubble within the column. Once the matrix is settled to give a column, it can be tested for presence of air channel and well packing by flowing a analyte with Kd=1, it is expected that the elution volume (V_e) in this case should be $V_o + V_i$.

2.   Sample Preparation: The sample is prepared in the mobile phase and it should be free of suspended particle to avoid clogging of the column. The most recommended method to apply the sample is to inject the sample with a syringe.

3.   Elution: In gel filtration column, no gradient of salt is used to elute the sample from the column. The flow of mobile phase is used to elute the molecules from the column.

4. Column Regeneration: After the analysis of analyte, gel filtration column is washed with the salt containing mobile phase to remove all non-specifically adsorb protein to the matrix. The column is then equiliberated with mobile phase to regenerate the column. The column can be store at 4 °C in the presence of 20% alcohol containing 0.05% sodium azide.

Determination of native molecular weight of a protein using gel filtration chromatography.

List of popular gel filtration matrix		
S.No.	Name of the matrix	Fractionation Range (Daltons)
1	Sephadex G10	Upto 700
2	Sephadex G25	1000-5000
3	Sephadex G50	1500-30,000
4	Sephadex G100	4000-150,000
5	Sephadex G200	5000-600,000
6	Sepharose 4B	60,000-20,000,000
7	Sepharose 6B	10,000-4,000,000

The molecular weight and size of a protein is related to the shape of the molecule and the relationship between molecular weight (M) and radius of gyration (R_g) is as follows:

$$R_g = \alpha M^a$$

Here "a" is a constant and it depends on shape of the molecule, a=1 for Rod, a=0.5 for coils and a=0.33 for spherical molecules.

The set of known molecular weight standard protein can be run on a gel filitration column and elution volume can be calculated from the chromatogram. A separate run with the analyte will give elution volume for unknown sample. Using following formula, Kd value for all standard protein and the test analyte can be calculated.

$$Kd = \frac{V_e - V_o}{V_i}$$

A plot of Kd versus log mol wt is given in figure, B and it will allow us to calculate the molecular weight of the unknown analyte.

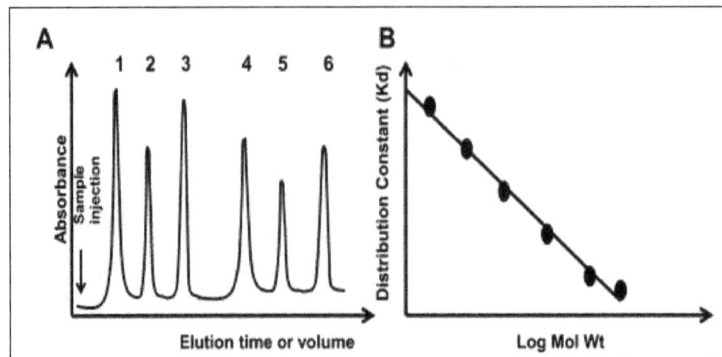

Determination of molecular weight by gel filtration chromatography. (A) Gel filtration chromatogram with the standard proteins (1-6), (B) Relationship between distribution constant (Kd) and Log Molecular weight.

Gel Filtration Chromatography-II

Oligomeric status of the protein-Native molecular weight determination by gel filtration in conjugation with the SDS-PAGE can be used to determine the oligomeric status of the protein.

$$\text{Oligomeric Status} = \frac{\text{Molecular weight (Gel Filtration)}}{\text{Molecular weight (SDS - PAGE)}}$$

Studying protein folding-Protein is madeup of the different types of amio acid residues linked by the peptide bond. As soon as peptide chain comes out from the ribosome, it folds into the 3-D conformation directed by the amino acid sequence, external environment and other factors. Protein structure has multilevel organization; Primary structure (sequence of protein), secondary (α-helix, β-sheet and turn), tertiary and quaternary structure. When protein is incubated with the increasing concentration of denaturing agents (such as urea), it unfolds the native structure into the unfolded extended conformation following multiple stages. The different protein conformation forms during unfolding pathway has distinct hydrodynamic surface area and it can be use to follow protein folding-unfolding stages with the gel filtration chromatography. The details of the experimental setup is given in the figure.

Protein is incubated with different concentration of urea (0-8M) for 8-10hrs at 370C. A gel filtration column is equiliberated with the buffer containing urea (same as in incubation mixture) and the incubation mixture is analyzed. As the concentration of denaturing agent is increasing, protein will unfold with an increase in hydrodynamic surface area. As a result, protein peak shifts towards left. At highest concentration of denaturant, protein unfolds completely and mostly appear in void volume.

Studying protein-ligand interaction-Gel filtration chromatography separates the molecules based on their size. Ligand binding to the protein induces conformational changes, result into the change in size or shape. In addition, ligand is small in size where as protein-ligand complex is big and may appear at a distinct place in the column. In step 1, a gel filtration column is equiliberated with the buffer and elution profile of ligand is recorded. Now column is equiliberated with the buffer containing ligand molecule. As the concentration of ligand is increased, protein binds ligand and form a larger complex with an increase in hydrodynamic surface area. As a result, protein peak shifts towards left.

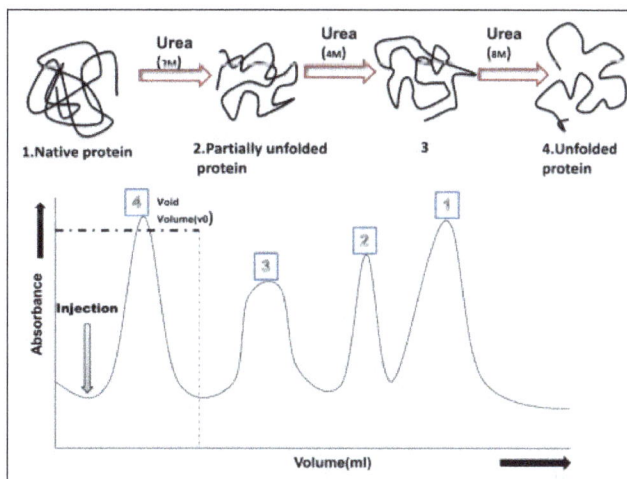

Studying Protein folding stages by gel filtration chromatography.

As the concentration of ligand will increase with a fixed amount of the protein, free ligand will appear in the chromatogram. The protein amount and the concentration at which free ligand appeared, and the elution data can be use to calculate the stoichiometric ratio of ligand/protein and the equilibrium constant.

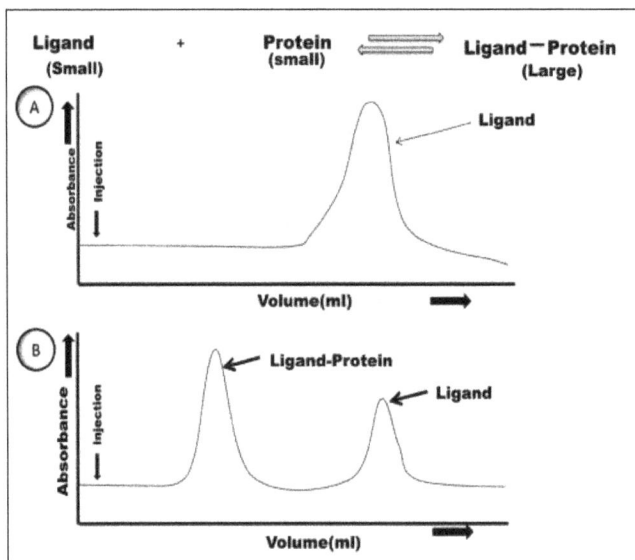

Studying Protein-ligand interaction by gel filtration chromatography.

Desalting- Desalting or removal of the small molecule from the protein is important for activity assay and other down stream processes. A gel filtration column is equilibrated with the buffer or water and then the sample for desalting is loaded. After the run the protein and salt are eluted separaetely as peak.

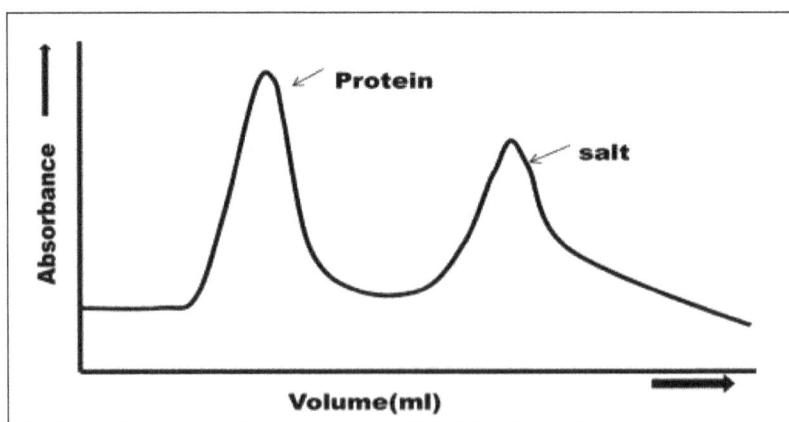

Desalting of a sample by gel filtration chromatography.

Affinity Chromatography-I

The chromatography techniques we discussed so far were exploiting different types of interactions between the matrix and the group present on the analyte but these chromatography techniques are not specific towards a particular analyte per se. The generalized chromatography approaches needs higher sample volume to isolate the molecule of interest. We will discuss another chromatography technique where a chromatography matrix is specific for a particular molecule or group of protein.

The affinity chromatography works on the principle of mutual recognition forces between a ligand and receptor. The major determinants, responsible to provide specificity are shape complementarity, electrostatic, hydrogen bonding, vander waal interaction between the groups present on the ligand-receptor pair. A mutual interaction between a ligand (L) and receptor (R) forms ligand-receptor complex (RL) with a dissociation constant Kd, which is expressed as follows:

$$R + L \rightleftharpoons RL$$

$$Kd = \frac{[R][L]}{[RL]}$$

Dissociation constant is specific to the receptor-ligand pair and number of interaction between them. when a crude mixture is passed through an affinity column, the receptor present on the matrix reacts with the ligand present on different molecules. The mutual collision between receptor on matrix and ligands from different molecule test the affinity between them and consequently the best choice bind to the receptor where as all other molecules do not bind and appear in flow through. A wash step removes remaining weakly bound molecules on matrix. Subsquently, a counter ligand is used to elute the bound molecule through a competition between the matrix bound molecule and counter ligand.

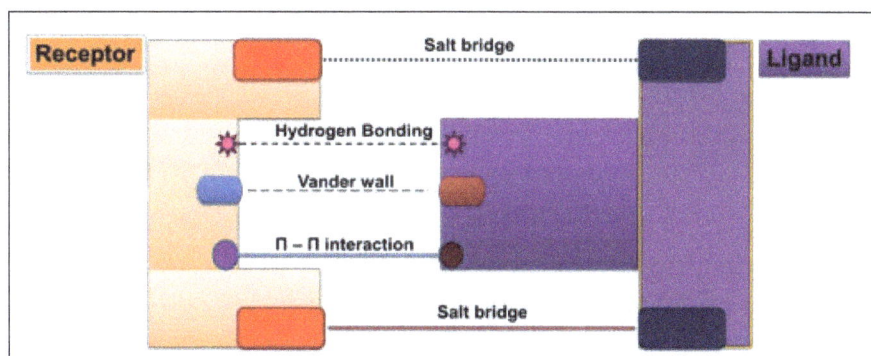

Interactions playing crucial role in providing specificity.

Advantages of Affinity Chromatography

1. Specificity: Affinity chromatography is specific to the analyte in comparison to other purification technique which are utilizing molecular size, charge, hydrophobic patches or isoelectric point etc.

2. Purification Yield: Compared to other purification method, affinity purification gives very high level of purification fold with high yield. In a typical affinity purification more than 90% recovery is possible.

3. Reproducible: Affinity purification is reproducible and gives consistent results from one purification to other as long as it is independent to the presence of contaminating species.

4. Easy to perform: Affinity purification is very robust and it depends on force governing ligand-receptor complex formation. Compared to other techniques, no column packing, no special purification system and sample preparation required for affinity purification.

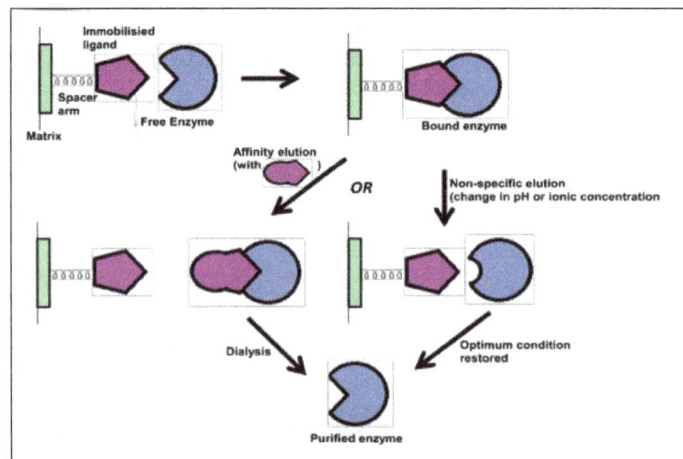

Principle of affinity chromatography.

- Different types of affinity chromatography: Affinity chromatography is further divided into the different types based on the nature of receptor present on matrix to binds tag present on the analyte molecule. Different types of affinity chromatography are:

- Bio-affinity chromatography: In this type of affinity chromatography, biomolecules are used as receptor present on matrix and it exploit the biological affinity phenomenon such as antibody-antigen. In addition, enzyme-substrate or enzyme- inhibitor is also belong to this class. Ex. GST-Glutathione.

- Pseudo-affinity chromatography: In this affinity chromatography, a non-biological molecule is used as receptor on matrix to exploit the separation and purification of biomolecules. There are two specific example to this class:

  ○ Dye-affinity chromatography: In this method, matrix is coupled to the reactive dye and the matrix bound dye has specificity towards a particular enzyme. For ex. Cibacron Blue F3G-A dye coupled to the dextran matrix has strong affinity towards dehydrogenases.

  ○ Metal-affinity chromatography: In this method, transition metals such as Fe^{2+}, Ni^{2+} or Zn^{2+} is coupled to the matrix and the matrix bound metal form multidentate complex with protein containing poly-his tag (6x His). The affinity of protein for matrix bound metal is different and these differences are been exploited in metal affinity chromatography to purify the protein.

- Covalent chromatography: This is a different type of chromatography technique where binding of analyte to the matrix is not reversible as it involves the formation a covalent bond between functional group present on matrix and analyte. Thiol group (- SH) present on neighbouring residues of protein forms disulphide bond after oxidation and under reducing environment, disulphide reversible broken back to free thiol group. The matrix in covalent chromatography has immobilized thio group which forms covalent linkage with the free thiol group containing protein present in the mixture. After a washing step to remove non-specifically bound protein, a mobile phase containing compound with reducing thio group is passed to elute the bound protein. The thio group containing compound present in mobile phase breaks the disulphide bond between protein and matrix thio group to release the protein in the mobile phase.

- Choice of matrix for Affinity chromatography: Different popular affinity matrix used for protein purification is given in table. The choice of matrix solely depends on the affinity tag present on the recombinant protein produced after genetic engineering.

S.No.	Receptor	Affinity towards protein ligand
	Matrix containing receptor for ligand present on protein.	
1	5'AMP	NAD^+-dependent dehydrogenase
2	2'5'-ADP	$NADP^+$-dependent dehydrogenase
3	Avidin	Biotin-containing enzymes
4	Protein A and Protein G	Immunoglobulin
5	Concanavalin A	Glycoprotein
6	Poly-A	Poly U mRNA
7	Lysine	rRNA
8	Cibacron Blue F3GA	NAD+ Containing dehydrogenase
9	Lectin	Glycoprotein
10	Heparin	DNA binding site

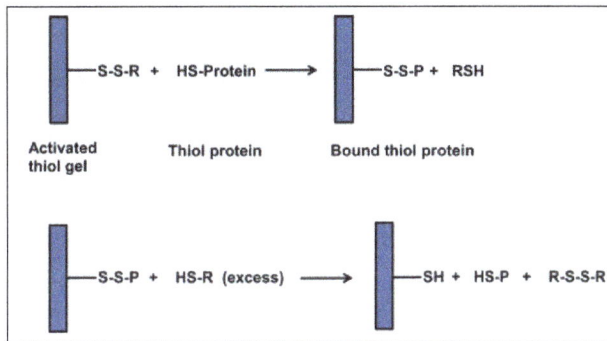

Principle of covalent chromatography.

Affinity Chromatography-II

- Generation of Receptor: The receptor molecule present on the matrix can be produced either by genetic engineering, isolation from the crude extract or in the case of antibody, it is produced in the mouse/rabbit model and purify. The generation of receptor molecule is beyond the discussion in the current lecture and interested student are advised to follow it from other relevant course.

- Coupling of the Receptor: Once the receptor molecule is available, it can be couple to the matrix by following steps. (1) Matrix activation (2) covalent coupling utilizing reactive group on ligand. (3) deactivation of the remaining active group on matrix.

- CNBr mediated receptor coupling: CNBr mediated coupling is more suitable for protein/peptide to the polysaccharide matrix such as agarose or dextran. CNBr reacts with polysaccharide at pH 11-12 to form reactive cynate ester with matrix or less reactive cyclic imidocarbonate group. Under alkaline condition these cynogen ester reacts with the amine group on receptor to form isourea derivative. The amount of cynate ester is more with agarose whereas imidacarbonate is more formed with dextran as a matrix. The protein or peptide ligand with free amino group is added to the activated matrix to couple the receptor for affinity purification.

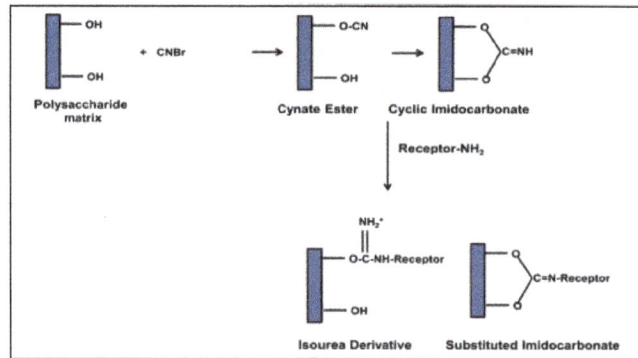

CNBr mediated coupling of receptor to the matrix.

Epichlorohydrin mediated receptor coupling-Epichlorohydrin activates the polysaccharide matrix by adding oxirane group with a 3 carbon alchol group (propanol) spacer arm. Activated matrix reacts with the receptors containing primary amine or thiol group. Receptor are couple to the matrix by a thioester or a secondary amine linkage. It can be able to couple hydroxyl group containing receptor molecule as well as by a ether linkage.

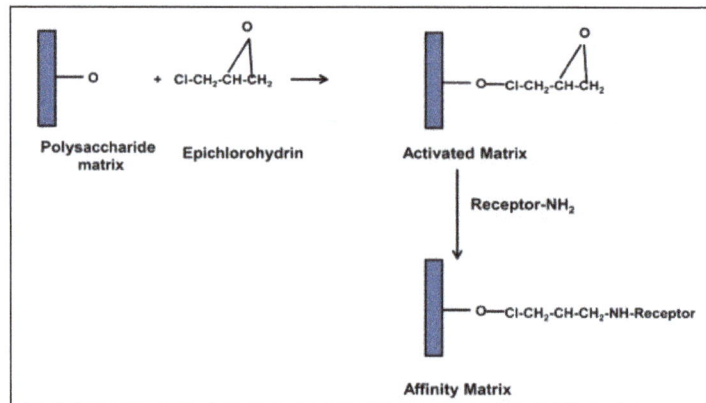

Epichlorohydrin mediated coupling of receptor to the matrix.

Carbodiimides mediated receptor coupling-Carboiimides reacts with the matrix containing carboxyl group to form isourea ester. The activated matrix is then allowed to react with the receptor molecule containing carboxyl or free amino group. Receptor are couple to the matrix by a secondary amine linkage.

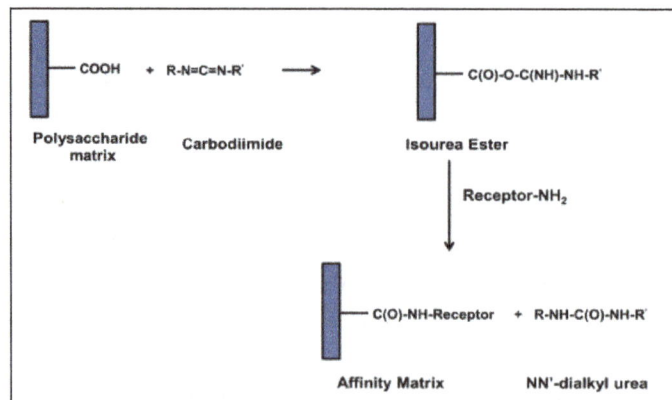

Carbodiimides mediated coupling of receptor to the matrix.

Operation of the Affinity chromatography: Different steps in affinity chromatography is given in figure.

1. Equlibration: Affinity column material packed in a column and equilibrate with a buffer containing high salt (0.5M NaCl) to reduce the non-specific interaction of protein with the analyte.

2. Sample Preparation: The sample is prepared in the mobile phase and it should be free of suspended particle to avoid clogging of the column. The most recommended method to apply the sample is to inject the sample with a syringe.

3. Elution: There are many ways to elute a analyte from the affinity column. (1) increasing concentration of counter ligand, (2) changing the pH polarity of the mobile phase, (3) By a detergent or chaotrophic salt to partially denature the receptor to reduce the affinity for bound ligand.

4. Column Regeneration: After the elution of analyte, affinity column requires a regeneration step to use next time. column is washed with 6M urea or guanidine hydrochloride to remove all non-specifically bound protein. The column is then equiliberated with mobile phase to regenerate the column. The column can be store at 4 °C in the presence of 20% alcohol containing 0.05% sodium azide.

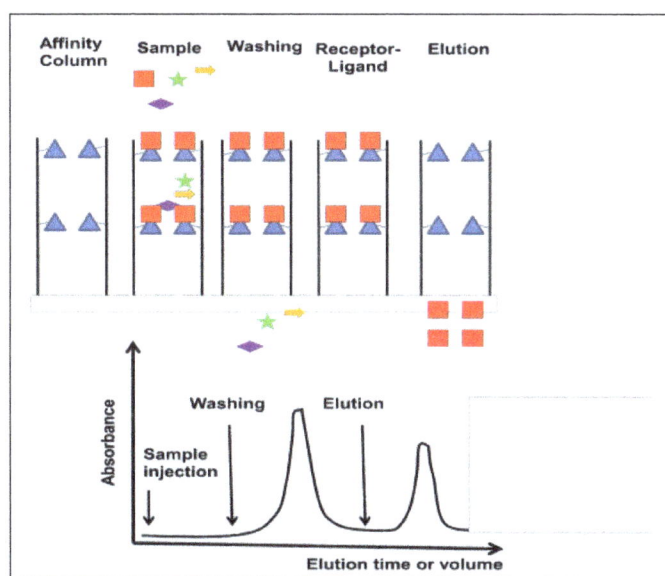

Performing Affinity chromatography.

Affinity Chromatography-III

Applications of Affinity Chromatography

1. Purification of biomolecules: Glutathione S-transferase (GST) utilizes glutathione as a substrate to catalyze conjugation reactions for xenobiotic detoxification purposes. The recombinant fusion protein contains GST as a tag is purified with glutathione coupled matrix. GST fusion protein is produced by the recombining protein of interest with the GST coding sequence present in the expression vector (either before or after coding sequence of protein of interest). It is transformed, over-expressed and the bacterial lysate containing fusion protein is purified, using affinity column. The sample is loaded on the column previously equiliberated with the buffer containing high salt (0.5M NaCl). Unbound protein is washed with the equilibration buffer and then the fusion proein

is eluted with different concentration of glutathione dissolved in the equilibration buffer. Purified fusion can be treated with the thrombin to remove the GST tag from the protein of interest. The mixture containing free GST tag and the protein can be purified using the affinity column again as tag will bind to the matrix but protein will come out in the unbound fraction.

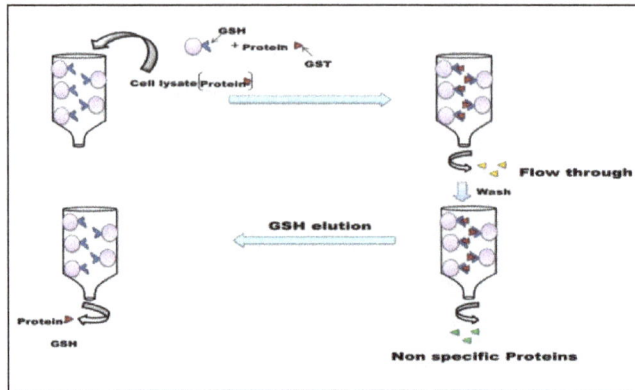

GST purification.

2. Protein-Protein interaction: Protein-protein interaction can be studied through multiple techniques or approaches. Affinity column also can be used as a tool to study or isolate interacting partner of a particular protein. A schematic figure to depict the steps involved in the studying protein-protein interaction is given in. In this approach, matrix is incubated with the pure protein-1 and then washed to ensure tight binding. All other sites on the bead is blocked with an non-specific protein such as BSA or an unrelated cell lysate. Now cell lysate or the pure protein-2 is passed to the protein-1 containing beads, followed by washing with the buffer to remove unbounded proteins. Now the protein-1 is eluted from the matrix either by adding high concentration of ligand, or with denaturing condition. Now the eluted protein is analyzed in the SDS-PAGE or SDS-PAGE followed by the western blotting to detect protein 1 or protein 2. As a control, cell lysate or protein-2 is also added to the matrix without protein-1 to rule out the possibility of protein-2 binding directly to the matrix.

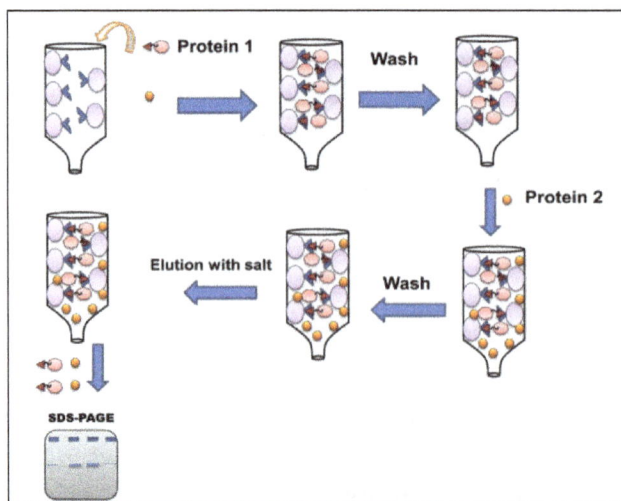

Protein-protein interaction studies with an affinity chromatography.

3. Enzymatic Assay: Affinity chromatography can be use to perform enzymatic assay such as protease assay. Peptides with different amino acid sequence bound with the terminal residue to the affinity

bead are incubated with the protease for an optimal time. Enzyme acts on the attached peptide and releases the free portion into the supernatant. The supernatant is recovered from the reaction mixture and can be analyzed in a MALDI-tof to deduce the amino acid sequence from the molecular weight. Analysis of a set of reaction may allow to predict the protease recognition and cutting site.

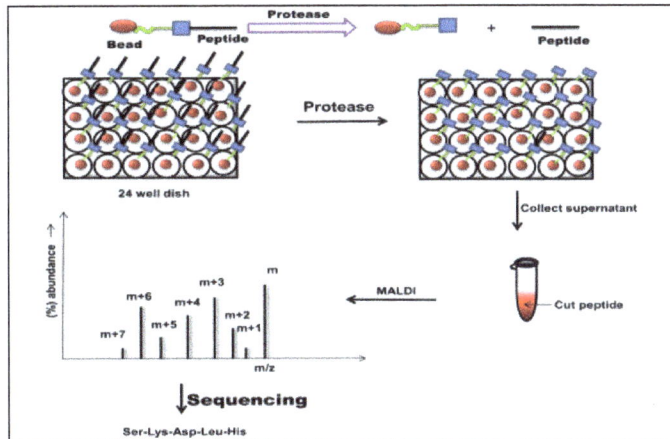

Protease assay by an affinity chromatography.

4. Clinical diagnosis: Receptor present on the matrix provides a unique tool to isolate, detect and characterize biomolecules from the crude mixture. For example, matrix containing boronic acid is used to separate and quantify glycosylated hemoglobin from diabetic patients blood. Ribonucleoside in patient urine can be identify by an affinity matrix containing boronic acid followed by the reverse phase chromatography.

5. Immuno-purification: The avidin-biotin system is used to capture and isolate cytokines from immune cells. Biotinylation of antibodies allows immobilization of antibodies in the correct orientation on the streptomycin coated glass beads. Lymphocyte lysate is passed to the column packed with the glass beads containing antibodies binds cytokines. The cytokines are eluted by flowing buffer of decreasing pH or by chaotrophic ions. The antibodies remain bound to the column due to strong affinity between avidin-biotin which is resistant to these chemical treatment.

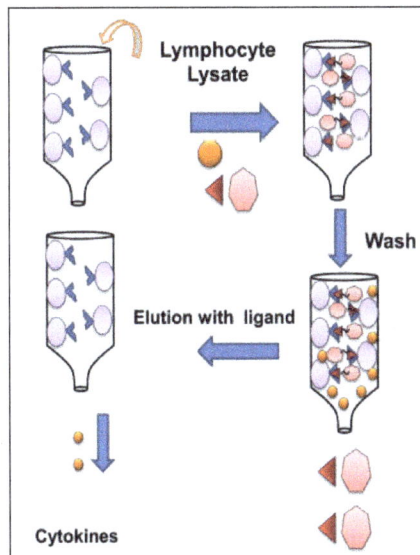

Immuno-purification with an affinity column.

Thin Layer Chromatography

The thin layer chromatography technique is an analytical chromatography to separate and analyze complex biological or non-biological samples into their constituents. It is most popular for monitoring the progress of a chemical reaction or estimation of a substance in a mixture. It is also one of the popular technique for testing the purity of a sample. In this method, the silica or alumina as a stationery phase is coated on to a glass or aluminium foil as thin layer and then a sample is allowed to run in the presence of a mobile phase (solvent). In comparison to other chromatography techniques, the mobile phase runs from bottom to top by diffusion (in most of the chromatography techniques, mobile phase runs from top to bottom by gravity or pump). As sample runs along with the mobile phase, it get distributed into the solvent phase and stationery phase. The interaction of sample with the stationery phase retard the movement of the molecule where as mobile phase implies an effective force onto the sample. Supose the force caused by mobile phase is F_m and the retardation force by stationery phase is F_s, then effective force on the molecule will be $(F_m - F_s)$ through which it will move. The molecule immobilizes on the silica gel (where, $F_m = F_s$) and the position will be controlled by multiple factors:

1. Nature or functional group present on the molecule or analyte.

2. Nature or composition of the mobile phase.

3. Thickness of the stationery phase.

4. Functional group present on stationery phase.

If the distance travelled by a molecule on TLC plate is D_m where as the distance travelled by the solvent is D_s, then the retardation factor (R_f) of molecule is given by:

$$R_f = \frac{\text{Distance travelled by substance } (D_m)}{\text{Distance travelled by solvent } (D_s)}$$

R_f value is characterstic to the molecule as long as the solvent system and TLC plate remains unchanged. It can be used to identify the substance in a crude mixture.

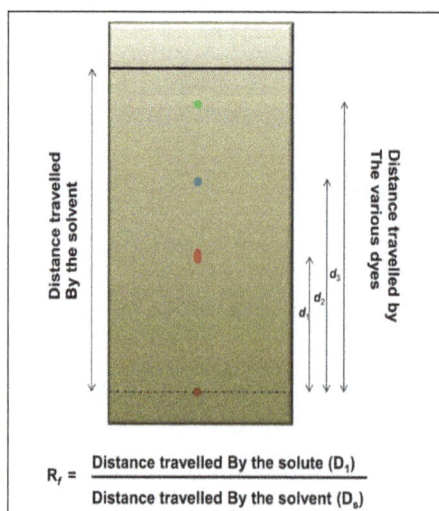

Principle of thin layer chromatography.

Operation of the Technique

Several steps are required to perform a thin layer chromatography to analyze a complex samples. These preparatory and operational steps are as follows:

1. Thin Layer Chromatography Chamber: Thin layer chromatographjy chamber (rectangular or cylindrical) is made up of transparent non-reactive material, mostly glass. It is covered from top with a thick glass sheet and the joints are sealed with a high vacuum grease to avoid loss of solvent vapor. All three sides of the chamber is covered with a whatman filter paper to uniformly equilibrate the chamber. A solvent system is filled in the chamber and it is allowed to humidify the chamber with the solvent vapor. It is important for uniform running of solvent front during TLC.

2. Preparation of TLC plate: A silica slurry is prepared in water and spread on the glass or alumina sheet as a thin layer and allowed to dry. It is baked at 110°C for 1hr in a hot air oven and then the plate is ready for TLC. The layer is thin ($\sim$ 0.1-0.25 mm) for analytical application and thick (0.4-2.1 mm) for preparative or bio-assay purposes.

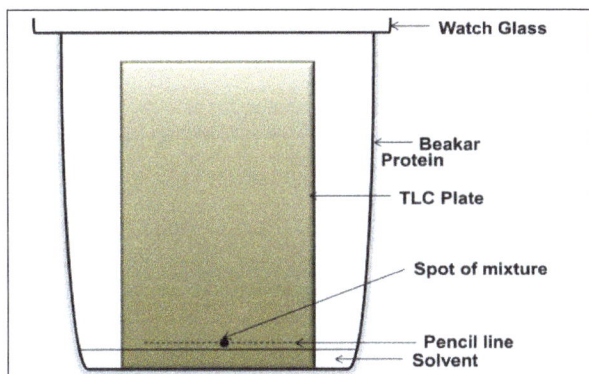

Thin Layer chromatography chamber.

3. Spoting: The events involved in spotting is given in figure. A line is draw with a pencil little away from the bottom. Sample is taken into the capillary tube or in a pipette. Capillary is touched onto the silica plate and sample is allowed to dispense. It is important that depending on the thickness of the layer, a suitable volume should be taken to apply. Spot is allowed to dry in air or a hair dryer can be used instead.

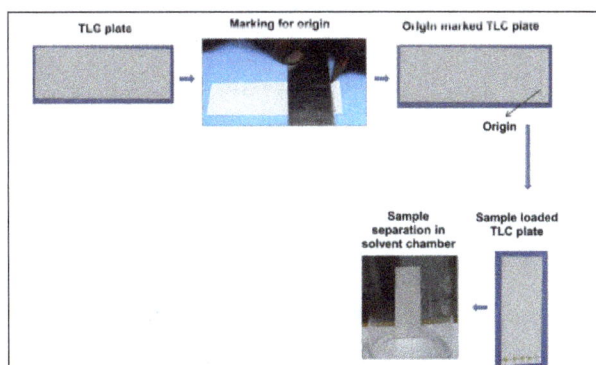

Events in spotting during thin layer chromatography.

4. Running of the TLC: Once the spot is dried, it is placed in the TLC chamber in such a way that spot should not be below the solvent level. Solvent front is allowed to move until the end of the plate.

Analysis of the Chromatography Plate

The plate is taken out from the chamber and air dried. If the compound is colored, it forms spot and for these substances there is no additional staining required. There are two methods of developing a chromatogram:

- Staining procedure: In the staining procedure, TLC plate is sprayed with the staining reagent to stain the functional group present in the compound. Forx. Ninhydrin is used to stain amino acids.

- Non-staining procedure: In non-staining procedure spot can be identify by following methods:

  ○ Autoradiography: A TLC plate can be placed along with the X-ray film for 48-72 hrs (exposure time depends on type and concentration of radioactivity) and then X-ray film is processed.

  ○ Fluorescence: Several heterocyclic compounds give fluorescence in UV due to presence of conjugate double bond system. TLC plate can be visualized in an UV- chamber to identify the spots on TLC plate.

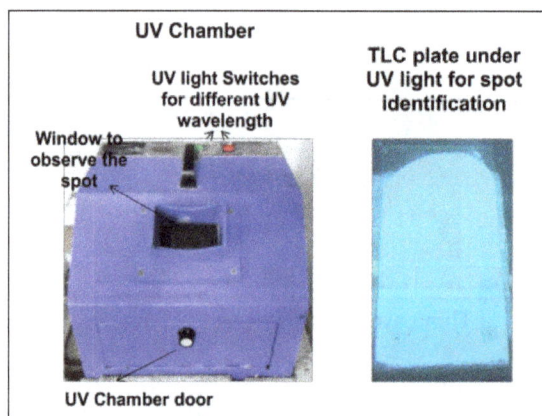

UV-Chamber and UV illuminated TLC plate.

Technical troubles with thin layer chromatography:

- Tailing effect: In general sample forms round circular spot on the TLC plate. It is due to the uniform movement of the solvent front through out the plate. But in few cases instead of forming a spot, a compound forms a spot with long trail or rocket shape spot it is due to few reasons as given below:

  ○ Over-loading: If the sample is loaded much more than the loading capacity of the TLC plate, it appears as spot with trail or rocket shape spot. A diluted sample can be tested to avoid this.

  ○ Fluctuation in temp or opening of chamber: If there will be fluntuation in temperature or solvent saturation in the chamber (due to opening of the chamber during running), it disturb the flow of solvent front and consequently cause spot with trail. it can be avoided by maintaining a uniform temperature and the opening of the chamber should be minimized especially during running.

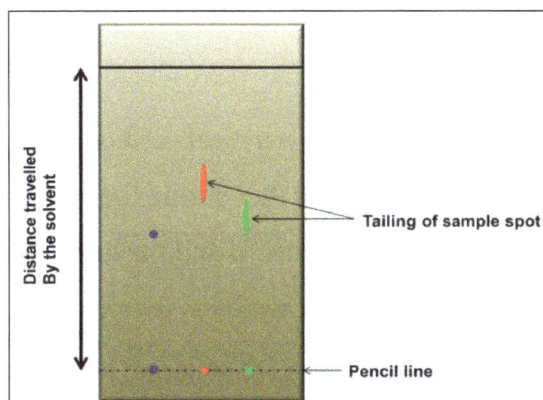

Tailing effect in thin layer chromatography.

- No movement of sample: In few cases, a sample doesn't move from the spot after the run is completed. These problems are common with high molecular weight substances such as protein or chemicals with large number of functional group. In this case, a change in polarity or pH of solvent system can be explored to bring the compound into the solvent front so that it run on silica plate to get resolved.

- Movement is too fast: In few cases, the movement of a compound is too fast and does not give time to interact with the matrix to resolve into individual compounds. In this case, a change in polarity of solvent system can be explored to retard the running of the sample.

Applications of Thin Layer Chromatography

- Composition analysis of biomolecules/synthetic preparation.

- Quality testing of compound.

- Identification of impurities in a sample.

- Progress of chemical reaction.

- Estimation of biomolecules.

- Bio-assay.

Chromatogram

A chromatogram is essentially the output of a chromatography run. It is an electronic file or hardcopy containing the information generated during the chromatography run.

There are many different variations on what is shown on a chromatogram — depending on the settings used in each laboratory and any regulatory requirements. As an example, the minimum shown on a GC run of an in-process sample might be:

- Sample identification (Product, batch number, stage number).

- Sample information (weight or concentration of sample).

- Date and time the injection was made.

- Analyst's name or identification.

- Instrument identification and name of analytical method used.

- Filename and location of raw data generated during the run.

- Chart recording showing the peaks generated and the baseline, known as a trace.

- Results table (containing raw data and calculated data).

Nowadays it is very much an automated process to generate a results table, but it wasn't always this simple.

Square Paper and Scissors

Before data analysis and digital integration become the standard: Graph paper and scissors were to be found on every analyst's shelf.

A chart recorder, linked to the detector, recorded the trace directly onto square paper or graph paper using an ink pen. From the trace, there were two methods commonly used for working out the results:

Counting Squares

The analyst used a ruler and pen to draw the best peak shape (triangle) and baseline. The number of squares in each triangle was then counted. This count was then used along with the detector's attenuation and the sample composition was calculated.

Cutting and Weighing

A baseline was constructed for the peaks that were recorded. The peaks were then cut out and weighed. The peak area being proportional to the weight, provided the paper's thickness and moisture content are uniform. Tricks like photocopying the trace and enlarging the trace could be used to increase precision.

As can be imaged, overlapping peaks and samples containing lots of different constituents could cause an analyst serious problems. Now, computers help reduce the errors.

Reading a Chromatogram

Although modern instrumentation has removed much of the guesswork from the paper and scissors days, the ability to read and interpret a chromatogram is just as important nowadays.

The trace should always be checked, as this will be the first indication if anything has gone wrong with the chromatography. Simple checks include:

- Does the baseline look OK, flat with not too much noise?

- Are the shapes of the peaks acceptable? Not too much asymmetry, nice and sharp and on-scale?

- Is there the right number of peaks? Any extra or missing peaks?

- Are the reference and internal standard peaks correctly identified, and the retention times OK?

Although the data is all recorded and calculations made for the analyst, it is the analyst's job to make sure that no errors have been made.

Chromatogram Evaluation

Electronic Image Acquisition

For electronic image acquisition the camera – like the human eye – captures polychromatic visible light. Under white light illumination it is the light reflected from the layer background. Under long-wavelength UV light (366 nm) it is the light emitted by fluorescent substances. When short-wavelength UV light (254 nm) is used, substances absorbing UV 254 nm appear as dark zones, provided the layer contains a fluorescence indicator (fluorescence quenching). The TLC Visualizer 2 is CAMAG's imaging and documentation system operated with vision-CATS. visionCATS enables an image-based evaluation of chromatograms for quantitative evaluation. The strength of the electronic image acquisition is the overview of the complete chromatogram.

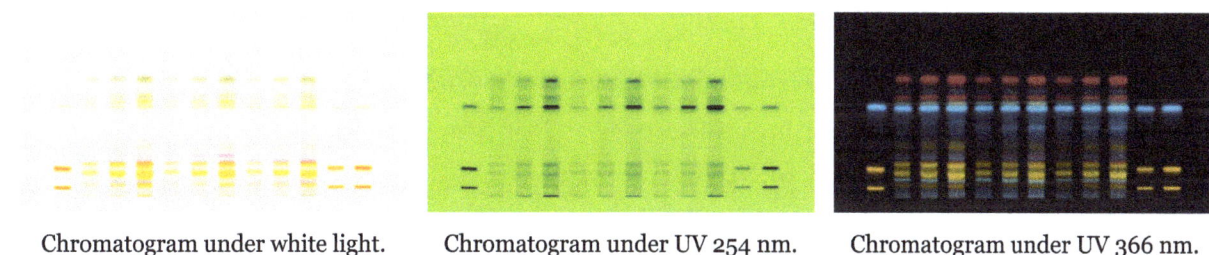

Chromatogram under white light.   Chromatogram under UV 254 nm.   Chromatogram under UV 366 nm.

Scanning Densitometry

In classical scanning densitometry the tracks of the chromatogram are scanned with monochromatic light in the form of a slit selectable in length and width. The spectral range of the CAMAG TLC Scanner 4 is 190–900 nm. Reflected light is measured either in the absorbance or in the fluorescence mode. From the acquired data quantitative results are computed with high precision and spectral selectivity.

With the TLC Scanner 4 absorption spectra can be recorded. The strengths of classical densitometry as compared with image evaluation are spectral selectivity and the higher precision of quantitative determinations.

Recommendations

Applying samples in the form of narrow bands allows densitometric evaluation by aliquot scanning, i.e. scanning with a slit about 2/3 of the track width. This improves reproducibility as the center portion of the sample zone is homogeneous and positioning errors, which can occur with samples applied as spots, are avoided.

For quantification sample zones should always be scanned with the wavelength of maximum absorbance which can be determined by spectra recording or by multi-wavelength scanning.

Retention Factor

A convenient way for chemists to report the results of a TLC plate in lab notebooks is through a "retention factor", or R_f value, which quantitates a compound's movement.

$$R_f = \frac{\text{Distance traveled by the compound}}{\text{Distance traveled by the solvent front}}$$

To measure how far a compound traveled, the distance is measured from the compound's original location (the baseline marked with pencil) to the compound's location after elution (the approximate middle of the spot). Due to the approximate nature of this measurement, ruler values should be recorded only to the nearest millimeter. To measure how far the solvent traveled, the distance is measured from the baseline to the solvent front.

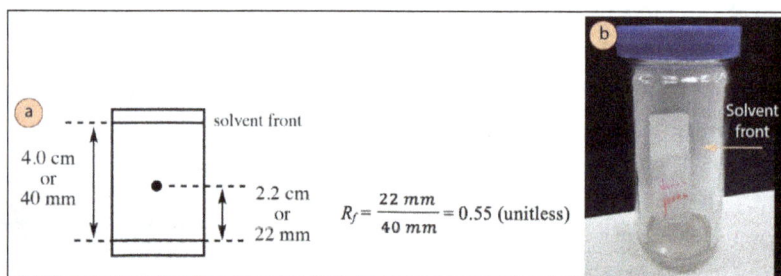

a) Sample R_f calculation, b) Appearance of the solvent front on an eluting TLC plate.

The solvent front is essential to this R_f calculation. When removing a TLC plate from its chamber, the solvent front needs to be marked immediately with pencil, as the solvent will often evaporate rapidly.

The R_f value is a ratio, and it represents the relative distance the spot traveled compared to the distance it could have traveled if it moved with the solvent front. An R_f of 0.55 means the spot moved 55% as far as the solvent front, or a little more than halfway.

Since an R_f is essentially a percentage, it is not particularly important to let a TLC run to any particular height on the TLC plate. In figure, a sample of acetophenone was eluted to different heights, and the R_f was calculated in each case to be similar, although not identical. Slight variations in R_f arise from error associated with ruler measurements, but also different quantities of adsorbed water on the TLC plates that alter the properties of the adsorbent. R_f values should always be regarded as approximate.

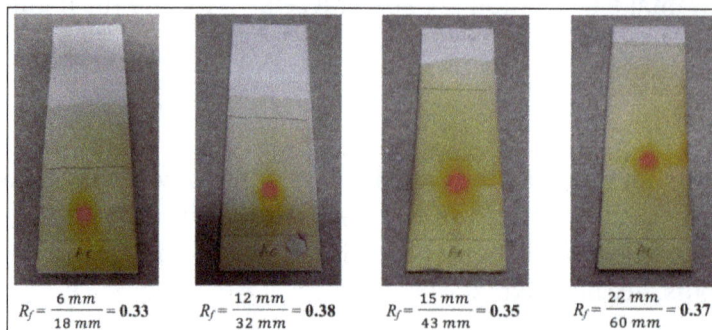

Acetophenone run to different heights on the TLC plate, using 6:1 hexanes: Ethyl acetate and visualized with anisaldehyde stain.

Although in theory a TLC can be run to any height, it's customary to let the solvent run approximately 0.5 cm from the top of the plate to minimize error in the R_f calculations, and to achieve the best separation of mixtures. A TLC plate should not be allowed to run completely to the top of the plate as it may affect the results. However, if using a saturated, sealed TLC chamber, the R_f can still be calculated.

Sometimes the R_f is called the retardation factor, as it is a measurement of how the movement of the spots is slowed, or retarded.

In paper chromatography, RF stands for retention factor, or the distance a liquid compound travels up a chromatography plate. The chromatography paper is the stationary phase and the liquid compound is the mobile phase; the liquid carries the sample solutions along the paper. When a liquid travels up the paper, it separates, allowing the person studying it to decipher the different components of the liquid solution. All compounds have a specific RF value for every specific solvent, and RF values are used to compare unknown samples with known compounds. Calculating RF is relatively simple with the right materials.

Calculating Retention Factor

Dip a strip of chromatography paper into the liquid solvent and the liquid solution to be analyzed. As the solvent is absorbed up the paper, the components of the solution will bleed out onto the paper.

Once the liquids have stopped moving, take the paper out of the liquid.

With your ruler, measure the distance the solvent traveled, which is D_f, and measure the distance the test solution traveled, which is D_s.

Calculate the retention factor using this equation: $R_f = D_s/D_f$. Simply divide the distance the solution traveled by the distance the solvent traveled. The retention factor will always be between zero and one. It cannot be zero because the substance must have moved, and it cannot be more than one because the solution cannot travel farther than the solvent.

Use the retention factor to compare to known retention factors and determine the substance with which you are working.

Full Proof Techniques of Chromatography

Partition Chromatography

Two immiscible liquid phases are passed through the supporting phase-containing materials. This is also called Liquid-Liquid Chromatography. Another type of Partition Chromatography is Gas-Liquid Partition Chromatography, where carrier gases such as N_2, helium, argon, H_2, and CO_2 are not carrier gases used as mobile phase.

Table: In Partition Chromatography one solvent acts as stationary and the other as mobile phase.

Stationary Phase	Mobile Phase
1. Water	1. Butanol
2. Aqueous acids	2. Pentanol

3. Aqueous alkalis	3. Benzyl alcohol
4. Buffer sol.	4. Phenol
5. Salf sol.	5. Benzene
6. Aq. Methanol etc.	6. Toluene etc.

The materials of Chromatographic phase are:

1. Silica gel,

2. Kieselguhr,

3. Cellulose,

4. Starch,

5. Dextrin, and

6. Rubber and Chlrorinated rubber.

Application

This method is used for the following analysis:

1. Amino acids and peptides separation;

2. Protein separation;

3. Separation of carbohydrates; ,

4. Organic acids;

5. Steroids.

Ion Exchange Chromatography

In this method ion exchange materials are used as the stationary phase in chromatography. A variety of materials have been used as ion exchanger.

Thus most important categories of ion exchangers are:

1. Natural and artificial siliceous minerals;

2. Sulphonated carbonaceous materials;

3. Synthetic polymers;

4. Derivatives of cellulose;

5. Miscellaneous.

Adsorption Chromatography

In this method certain adsorbents are used for separation technique. Many substances of widely different chemical nature have been used as adsorbents.

The essential apparatus for a chromatographic separation is very simple. It consists of a tube of suitable dimensions, with means for supporting the packed adsorbent and openings for the admission and collection of mobile phase. Recently, multiple coupled filter column is used for this purpose.

The major types of adsorbents are:

a. Polar: Alumina (Al_2IO_3), $CaCO_3$, Cellulose, Kaolin, MgO, Magnesium Silicate, Silicic acid;

b. Nonpolar: Carbon.

Application

It is used for analysis of porphyrins separation.

Gel Permeation Chromatography

Gel Permeation Chromatography (GPC) is a liquid chromatographic method known variously as gel filtration and Molecular Exclusion Chromatography (MEC). The gel structure contains pores of varying diameters, up to a maximum size.

The test molecules are washed through a column of the gel and molecules larger than the largest pores in the gel are excluded from the gel structure. Smaller molecules, however, penetrate the gel to a varying extent depending upon their size and this retards their progress through the column.

Elution, therefore, is in order of decreasing size. There are a variety of products available for Gel Permeation Chromatography and they are usually classified according to their exclusion limit, i.e., the relative molecular mass above which all molecules are excluded from the gel structure.

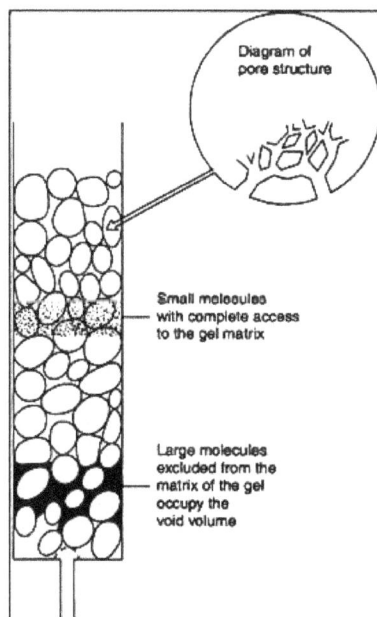

Separation by Gel Permeation Chromatography.

Sephadex, the original medium, is based on dextran (a linear glucose polymer) that is modified to give varying degrees of cross-linking which determines the pore size of the material. It is a strongly hydrophilic polymer and swells considerably in water; before a column is prepared, the gel must be fully hydrated.

Polyacrylamide beads are also available with a wide range of pore sizes and are prepared commercially in a manner similar to that described for polyacrylamide electrophoresis. They are also hydrophilic and swell significantly in aqueous solutions but are chemically more stable than the dextran gels. Agarose gels are particularly useful when gels with a very large pore size are required.

They are also hydrophilic but are usually sold in the swollen form. A major problem is the fact that they soften at temperatures above 30 °C.

Mixed gels of polyacrylamide and agarose are available (Ultra-gel, LKB) in which the polyacrylamide provides a three-dimensional structure which supports the interstitial agarose gel. Polystyrene gels are hydrophobic and, as a result, are used primarily with non-aqueous solvents and for organic chemical applications rather than with biological samples.

Table : List of Gel Permeation Media.

Trade Name and Manufacturer	Chemical Nature	Fractionation Range (RMM)
Pharmacia		
Sephadex G 10 to G 200	Cross-linked dextrans	Up to 700 5,000-800,000
Sepharose 6 B to 2 B	Agarose	Up to 4×10^4
Sepharose C1-6 B to C1-2B		Up to 4×10^4 Up to 40×10^4
Biored		
Biogel P2 to P300	Cross-linked polyacrylamide	200-2,500 100,000-400,000
Biogel A 0.5 to A 150	Agarose	10,000-500,000 1×10^5 -150×10^4
Biobeads S × 1 to S × 12	Polystyrene	600-14,000 Up to 400
LKB		
Ultragel AcA22 to AcA54	Agrose and polyacrylamide	100,000-1.2×10^4 5,000-70,000
RMM = Relative Molecular Mass		

Gas Chromatography Principle

Gas-Liquid Chromatography (GLC) accomplishes a separation by partitioning solutes between a mobile gas phase and a stationary liquid phase held on a solid support. Gas-Solid Chromatography (GSC) employs a solid adsorbent as the stationary phase.

The sequence of a gas chromatographic separation is:

A sample containing the solutes is injected into a heating block where it is immediately vaporized and swept as a plug of vapour by the carrier gas stream into the column inlet. The solutes are adsorbed at the head of the column by the stationary phase and then desorbed by fresh carrier gas.

This sorption- desorption process occurs repeatedly as the sample is moved toward the column outlet by the carrier (mobile phase) gas. Each solute will travel at its own rate through the column. Their bands will separate to a degree that is determined by the individual partition ratios and the extent of band spreading.

The solutes are eluted sequentially in the increasing order of their partition ratios and enter a detector attached to the column exit. If a recorder is used, the signals appear on the chart as a plot of time versus the composition of the carrier gas stream. The time of emergence of a peak is characteristic for each component; the peak 'area is proportional to the concentration of the component in the mixture.

Although Gas Chromatography is limited to volatile materials — about 15% of all organic compounds — the availability of column temperatures up to 450°C, pyro lytic techniques, and the possibility of converting non-volatile materials into a volatile derivative extend somewhat the applicability of the method.

Basically, a Gas Chromatograph consists of six parts:

1. A supply of carrier gas in a high pressure cylinder with attendant pressure regulators and flow meters, and a valve to introduce extra make-up gas to some detectors,

2. A sample injection system,

3. The separation column,

4. The detector,

5. An electrometer and strip-chart recorder (and integrator perhaps), and

6. Separate thermo-stated compartments for housing the column and the detector so as to regulate their temperature, or to program the column temperature.

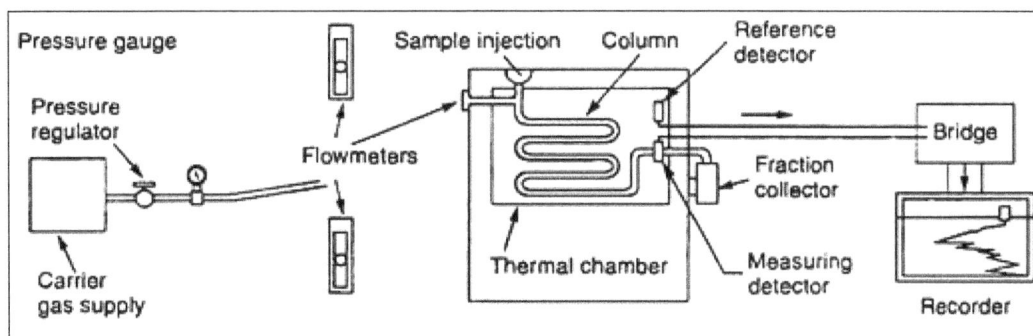

Schematic view of a Gas Chromatograph.

The most exacting problem in Gas Chromatography is presented by the sample injection system. The sample must be introduced as a vapour in the smallest possible volume and in a minimum of time without decomposition or fraction occurring. Generally, two basic type of columns, viz., packed or capillary columns, are used for analytical purpose.

A variety of detector systems is used for analytical purpose, viz., Photoionization (PID), Flame ionization (FID), Electron capture (ECD), Thermal conductivity (TCD) and Flame photometric (FPD).

Among them, thermal conductivity detectors were first and are still widely used; their simplicity is an advantage and they are non-destructive. In contrast — for high-sensitivity analyses of organic compounds — the hydrogen flame ionization detector is used.

High Performance Liquid Chromatography (HPLC)

This is a specialized version of Column Chromatography, where much more precision analysis of organics will be available. The wide applicability, speed and sensitivity of HPLC have resulted in it becoming the most popular form of chromatography and virtually all types of biological molecules have been analyse or purified using this technique.

There are four distinct modes of HPLC separation methods — Adsorption, Partition, Ion Exchange and Exclusion. All of them vary with the nature of stationary phase.

Some examples of HPLC stationary phases are given in table.

Chromatographic Separation Principle	Commercial Name	Nature of Stationary Phase	Type of Support
Adsorption	Partisil C_8	Octylsilane	Porus
	Corasil	Silica	Pellicular
	Pellumina	Alumina	Pellicular
	Parisil	Silica	Microporous
	Micropak A1	Alumina	Microporous
Partition	Bondapak-C_{18}/Corasil	Octadecylsilane (ODS)	Pellicular
	ì Bondapak-C_{18}	Octadecylsilane	Porous
	ULTRApak TSK ODS	Octadecylsilane	Porous
	ì Bondapak-NH_2	Alkylamine	Porous
	ULTRApak TSK-NH_2	Alkylamine	Porous
Ion-exchange	Partisil SAX	Strong base	Porous
	MicroPak-NH_2	Weak base	Porous
	Partisil-SCX	Strong acid	Porous
	AS Pellionex-SAX	Strong base	Pellicular
	Zipak-WAX	Weak base	Pellicular
	Peisorb-KAT	Strong acid	Pellicular
Exclusion	BioGlas	Glass	Rigid solid
	Styragel	Polystyrene-divinylbenzene	Semi-rigid gel
	Superose	Agarose	Soft gel
	Fractoge TSK	Polyvinylchloride	Semi-rigid gel

Chromatography Detector

A chromatography detector is a device used in gas chromatography (GC) or liquid chromatography (LC) to detect components of the mixture being eluted off the chromatography column. There are two general types of detectors: destructive and non-destructive. The destructive detectors perform continuous transformation of the column effluent (burning, evaporation or mixing with reagents) with subsequent measurement of some physical property of the resulting material (plasma, aerosol or reaction mixture). The non-destructive detectors are directly measuring some property of the column eluent (for example UV absorption) and thus affords greater analyte recovery.

Destructive Detectors

In liquid chromatography:

- Charged aerosol detector (CAD).

- Evaporative light scattering detector (ELSD).

In gas chromatography:

- Flame ionization detector (FID).

- Flame photometric detector (FPD).

- Nitrogen Phosphorus Detector (NPD).

- Atomic-emission detector (AED).

In all types of chromatography:

- Mass spectrometer (MS).

Non-destructive Detectors

In liquid chromatography:

- UV detectors, fixed or variable wavelength, which includes diode array detector (DAD or PDA): The UV absorption of the effluent is continuously measured at single or multiple wavelengths. These are by far most popular detectors for LC.

- Fluorescence detector: Irradiates the effluent with a light of set wavelength and measure the fluorescence of the effluent at a single or multiple wavelength.

- Refractive index detector (RI or RID): Continuously measures the refractive index of the effluent. The lowest sensitivity of all detectors. Often used in size exclusion chromatography for polymer analysis.

- Radio flow detector: Measures radioactivity of the effluent. This detector can be destructive if a scintillation cocktail is continuously added to the effluent.

- Chiral detector continuously measures the optical angle of rotation of the effluent: It is used only when chiral compounds are being analyzed.

- Conductivity monitor: Continuously measures the conductivity of the effluent. Used only when conductive eluents (water or alcohols) are used.

In gas chromatography:

- Thermal conductivity detector, (TCD): Measures the thermal conductivity of the eluent.

- Electron capture detector, (ECD): The most sensitive detector known. Allows for the detection of organic molecules containing halogen, nitro groups etc.

- Photoionization detector, (PID): Measures the increase in conductivity achieved by ionizing the effluent gas with UV radiation.

References

- Chromatography, science: britannica.com, Retrieved 29 April, 2019

- Top-12-types-of-chromatographic-techniques-biochemistry, chromatography-techniques, biochemistry: biologydiscussion.com, Retrieved 30 June, 2019

- What-is-a-chromatogram, breaking-news, hplc-uhplc, news: chromatographytoday.com, Retrieved 19 August, 2019

- Evaluation-documentation-tlc-ms-bioluminenscence, products, tlc-hptlc: camag.com, Retrieved 10 May, 2019

- How-7152385-calculate-rf: sciencing.com, Retrieved 25 May, 2019

- 6-full-proof-techniques-of-chromatography, chromatography, micro-biology: biologydiscussion.com, Retrieved 19 July, 2019

Gas Chromatography 2

- **GC Detectors**
- **Principle of Gas Chromatography**
- **Physical Components of GC**
- **Gas Chromatography–Vacuum Ultraviolet Spectroscopy**
- **Headspace Gas Chromatography**
- **Preparative Gas Chromatography**
- **Disadvantages and Advantages of GC**

Gas chromatography refers to a common type of chromatography which is used to separate and analyze compounds which can be vaporized without decomposition. This chapter has been carefully written to provide an easy understanding of gas chromatography including its principles, physical components, disadvantages and advantages.

Gas chromatography is a term used to describe the group of analytical separation techniques used to analyze volatile substances in the gas phase. In gas chromatography, the components of a sample are dissolved in a solvent and vaporized in order to separate the analytes by distributing the sample between two phases: a stationary phase and a mobile phase. The mobile phase is a chemically inert gas that serves to carry the molecules of the analyte through the heated column. Gas chromatography is one of the sole forms of chromatography that does not utilize the mobile phase for interacting with the analyte. The stationary phase is either a solid adsorbant, termed gas-solid chromatography (GSC), or a liquid on an inert support, termed gas-liquid chromatography (GLC).

In early 1900s, Gas chromatography (GC) was discovered by Mikhail Semenovich Tsvett as a separation technique to separate compounds. In organic chemistry, liquid-solid column chromatography is often used to separate organic compounds in solution. Among the various types of gas chromatography, gas-liquid chromatography is the method most commonly used to separate organic compounds. The combination of gas chromatography and mass spectrometry is an invaluable tool in the identification of molecules. A typical gas chromatograph consists of an injection port, a column, carrier gas flow control equipment, ovens and heaters for maintaining temperatures of the injection port and the column, an integrator chart recorder and a detector.

To separate the compounds in gas-liquid chromatography, a solution sample that contains organic compounds of interest is injected into the sample port where it will be vaporized. The vaporized samples that are injected are then carried by an inert gas, which is often used by helium or nitrogen. This inert gas goes through a glass column packed with silica that is coated with a liquid. Materials that are less soluble in the liquid will increase the result faster than the material with greater solubility.The purpose of this module is to provide a better understanding on its separation and measurement techniques and its application.

In GLC, the liquid stationary phase is adsorbed onto a solid inert packing or immobilized on the capillary tubing walls. The column is considered packed if the glass or metal column tubing is packed with small spherical inert supports. The liquid phase adsorbs onto the surface of these beads in a thin layer. In a capillary column, the tubing walls are coated with the stationary phase or an adsorbant layer, which is capable of supporting the liquid phase. However, the method of GSC, has limited application in the laboratory and is rarely used due to severe peak tailing and the semi-permanent retention of polar compounds within the column. Therefore, the method of gas-liquid chromatography is simply shortened to gas chromatography and will be referred to as such here. The purpose of this module is to provide a better understanding on its separation and measurement techniques and its application.

Instrumentation

Sample Injection

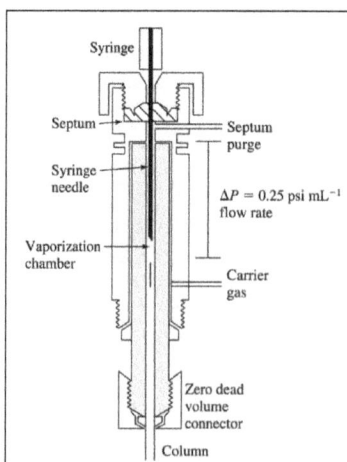

A cross-sectional view of a microflash vaporizer direct injector.

A sample port is necessary for introducing the sample at the head of the column. Modern injection techniques often employ the use of heated sample ports through which the sample can be injected and vaporized in a near simultaneous fashion. A calibrated microsyringe is used to deliver a sample volume in the range of a few microliters through a rubber septum and into the vaporization chamber. Most separations require only a small fraction of the initial sample volume and a sample splitter is used to direct excess sample to waste. Commercial gas chromatographs often allow for both split and splitless injections when alternating between packed columns and capillary columns. The vaporization chamber is typically heated 50 °C above the lowest boiling point of the sample and subsequently mixed with the carrier gas to transport the sample into the column.

Carrier Gas

The carrier gas plays an important role, and varies in the GC used. Carrier gas must be dry, free of oxygen and chemically inert mobile-phase employed in gas chromatography. Helium is most commonly used because it is safer than, but comprable to hydrogen in efficiency, has a larger range of flow rates and is compatible with many detectors. Nitrogen, argon, and hydrogen are also used depending upon the desired performance and the detector being used. Both hydrogen and helium, which are commonly used on most traditional detectors such as Flame Ionization(FID), thermal conductivity (TCD) and Electron capture (ECD), provide a shorter analysis time and lower elution temperatures of the sample due to higher flow rates and low molecular weight. For instance, hydrogen or helium as the carrier gas gives the highest sensitivity with TCD because the difference in thermal conductivity between the organic vapor and hydrogen/helium is greater than other carrier gas. Other detectors such as mass spectroscopy, uses nitrogen or argon which has a much better advantage than hydrogen or helium due to their higher molecular weights, in which improve vacuum pump efficiency.

All carrier gasses are available in pressurized tanks and pressure regulators, gages and flow meters are used to meticulously control the flow rate of the gas. Most gas supplies used should fall between 99.995% - 99.9995% purity range and contain a low levels (< 0.5 ppm) of oxygen and total hydrocarbons in the tank. The carrier gas system contains a molecular sieve to remove water and other impurities. Traps are another option to keep the system pure and optimum sensitive and removal traces of water and other contaminants. A two stage pressure regulation is required to use to minimize the pressure surges and to monitor the flow rate of the gas. To monitor the flow rate of the gas a flow or pressure regulator was also require onto both tank and chromatograph gas inlet. This applies different gas type will use different type of regulator. The carrier gas is preheated and filtered with a molecular sieve to remove impurities and water prior to being introduced to the vaporization chamber. A carrier gas is typically required in GC system to flow through the injector and push the gaseous components of the sample onto the GC column, which leads to the detector.

Table: Gas Recommendations for Capillary Columns.

Detector	Carrier gas	Preferred makeup gas	Second choice	Detector, anode purge, or reference gas
Electron Capture	Hydrogen	Argon/Methane	Nitrogen	A node purge must be same as makeup
	Helium	Argon/Methane	Nitrogen	
	Nitrogen	Nitrogen	Argon/Methane	

	Argon/Methane	Argon/Methane	Nitrogen	
Flame Ionization	Hydrogen	Nitrogen	Helium	Hydrogen and air for detector
	Helium	Nitrogen	Helium	
	Nitrogen	Nitrogen	Helium	
Flame Photometric	Hydrogen	Nitrogen		Hydrogen and air for detector
	Helium	Nitrogen		
	Nitrogen	Nitrogen		
	Argon	Nitrogen		
Nitrogen Phosphorus	Helium	Nitrogen	Helium	Hydrogen and air for detector
	Nitrogen	Nitrogen	Helium	
Thermal Conductivity	Hydrogen	Must be same as carrier and reference gas	Must be same as carrier and reference gas	Reference must be same as carrier and makeup
	Helium			
	Nitrogen			

Table: Gas Recommendations for Packed Columns.

Detector	Carrier gas	Comments	Detector, anode purge or reference gas
Electron Capture	Nitrogen	Maximum sensitivity	Nitrogen
	Argon/Methane	Maximum dynamic range	Argon/Methane
Flame Ionization	Nitrogen	Maximum sensitivity	Hydrogen and air for detector
	Helium	Acceptable alternative	
Flame photometric	hydrogen		Hydrogen and air detector
	Helium		
	Nitrogen		
	Argon		
Nitrogen Phosphorus	Helium	Optimum performance	Hydrogen and air for detector
	Nitrogen	Acceptable alternative	
Thermal Conductivity	Helium	General use	Reference must be same as carrier
	Hydrogen	Maximum sensitivity	
	Nitrogen	Hydrogen detection	
	Argon	Maximum hydrogen sensitivity	

Column Oven

The thermostatted oven serves to control the temperature of the column within a few tenths of a degree to conduct precise work. The oven can be operated in two manners: isothermal programming or temperature programming. In isothermal programming, the temperature of the column is held constant throughout the entire separation. The optimum column temperature for isothermal operation is about the middle point of the boiling range of the sample. However, isothermal programming works best only if the boiling point range of the sample is narrow. If a low isothermal column temperature is used with a wide boiling point range, the low boiling fractions are well resolved but the high boiling fractions are slow to elute with extensive band broadening. If the temperature is increased closer to the boiling points of the higher boiling components, the higher boiling components elute as sharp peaks but the lower boiling components elute so quickly there is no separation.

In the temperature programming method, the column temperature is either increased continuously or in steps as the separation progresses. This method is well suited to separating a mixture with a broad boiling point range. The analysis begins at a low temperature to resolve the low boiling components and increases during the separation to resolve the less volatile, high boiling components of the sample. Rates of 5-7 °C/minute are typical for temperature programming separations.

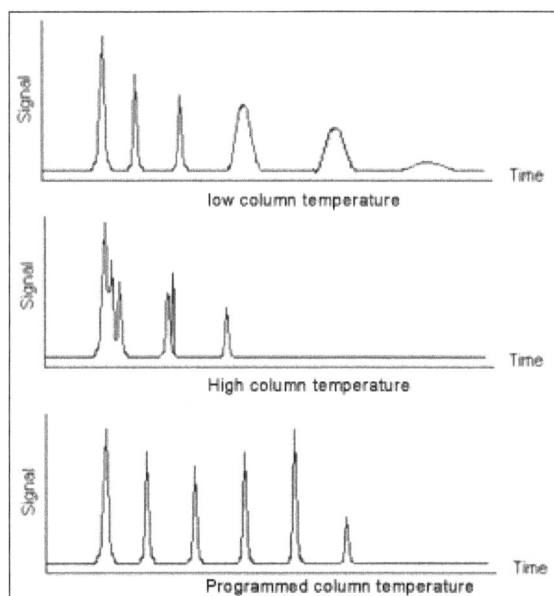

The effect of column temperature on the shape of the peaks.

Open Tubular Columns and Packed Columns

Open tubular columns, which are also known as capillary columns, come in two basic forms. The first is a wall-coated open tubular (WCOT) column and the second type is a support-coated open tubular (SCOT) column. WCOT columns are capillary tubes that have a thin later of the stationary phase coated along the column walls. In SCOT columns, the column walls are first coated with a thin layer (about 30 micrometers thick) of adsorbant solid, such as diatomaceous earth, a material which consists of single-celled, sea-plant skeletons. The adsorbant solid is then treated with the liquid stationary phase. While SCOT columns are capable of holding a greater volume of stationary phase than a WCOT column due to its greater sample capacity, WCOT columns still have greater column efficiencies.

Most modern WCOT columns are made of glass, but T316 stainless steel, aluminum, copper and plastics have also been used. Each material has its own relative merits depending upon the application.

Glass WCOT columns have the distinct advantage of chemical etching, which is usually achieved by gaseous or concentrated hydrochloric acid treatment. The etching process gives the glass a rough surface and allows the bonded stationary phase to adhere more tightly to the column surface.

One of the most popular types of capillary columns is a special WCOT column called the fused-silica wall-coated (FSWC) open tubular column. The walls of the fused-silica columns are drawn from purified silica containing minimal metal oxides. These columns are much thinner than glass columns, with diameters as small as 0.1 mm and lengths as long as 100 m. To protect the column, a polyimide coating is applied to the outside of the tubing and bent into coils to fit inside the thermostatted oven of the gas chromatography unit. The FSWC columns are commercially available and currently replacing older columns due to increased chemical inertness, greater column efficiency and smaller sampling size requirements. It is possible to achieve up to 400,000 theoretical plates with a 100 m WCOT column, yet the world record for the largest number of theoretical plates is over 2 million plates for 1.3 km section of column.

Packed columns are made of a glass or a metal tubing which is densely packed with a solid support like diatomaceous earth. Due to the difficulty of packing the tubing uniformly, these types of columns have a larger diameter than open tubular columns and have a limited range of length. As a result, packed columns can only achieve about 50% of the efficiency of a comparable WCOT column. Furthermore, the diatomaceous earth packing is deactivated over time due to the semi-permanent adsorption of impurities within the column. In contrast, FSWC open tubular columns are manufactured to be virtually free of these adsorption problems.

Table: Properties of gas chromatography columns.

	Types of Column			
	FSWC	WCOT	SCOT	Packed
Length	10 to 1000 m	10 to 1000 m	10 to 100 m	1 to 6 m
Inner Diameter	0.1 to 0.3 mm	0.25 to 0.75 mm	0.5 mm	2 to 4 mm
Efficiency (plate/m)	2000 to 4000	1000 to 4000	600 to 1200	500 to 1000
Sample Size	10 to 75 ng	10 to 1000 ng	10 to 1000 ng	10 to 10^6 ng
Pressure	Low	Low	Low	High
Speed	Fast	Fast	Fast	Slow
Inertness	Best	Good	Fair	Poor

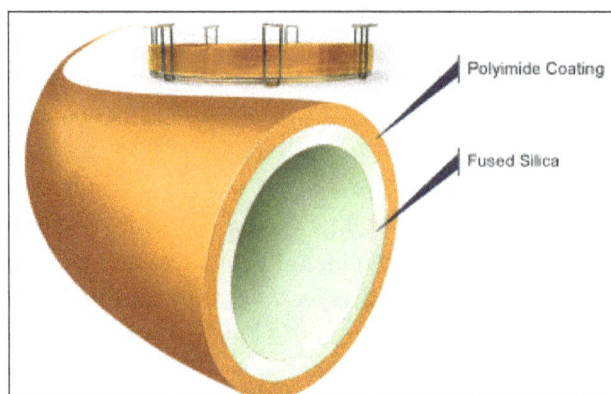

Computer Generated Image of a FSWC column (specialized for measuring BAC levels).

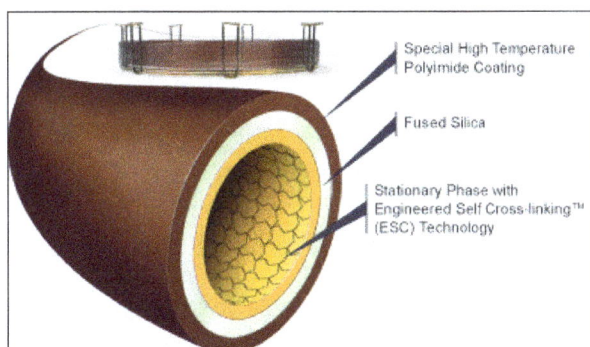

Computer Generated Image of a FSWC column (specialized to withstand extreme heat).

Different types of columns can be applied for different fields. Depending on the type of sample, some GC columns are better than the others. For example, the FSWC column shown in figure is designed specially for blood alcohol analysis. It produces fast run times with baseline resolution of key components in under 3 minutes. Moreover, it displays enhanced resolutions of ethanol and acetone peaks, which helps with determining the BAC levels. This particular column is known as Zebron-BAC and it made with polyimide coating on the outside and the inner layer is made of fused silica and the inner diameter ranges from .18 mm to .25 mm. There are also many other Zebron brand columns designed for other purposes.

Another example of a Zebron GC column is known as the Zebron-inferno. Its outer layer is coated with a special type of polyimide that is designed to withstand high temperatures. As shown in figure, it contains an extra layer inside. It can withstand up to 430 °C to be exact and it is designed to provide true boiling point separation of hydrocarbons distillation methods. Moreover, it is also used for acidic and basic samples.

Detection Systems

The detector is the device located at the end of the column which provides a quantitative measurement of the components of the mixture as they elute in combination with the carrier gas. In theory, any property of the gaseous mixture that is different from the carrier gas can be used as a detection method. These detection properties fall into two categories: bulk properties and specific properties. Bulk properties, which are also known as general properties, are properties that both the carrier gas and analyte possess but to different degrees. Specific properties, such as detectors that measure nitrogen-phosphorous content, have limited applications but compensate for this by their increased sensitivity.

Each detector has two main parts that when used together they serve as transducers to convert the detected property changes into an electrical signal that is recorded as a chromatogram. The first part of the detector is the sensor which is placed as close the the column exit as possible in order to optimize detection. The second is the electronic equipment used to digitize the analog signal so that a computer may analyze the acquired chromatogram. The sooner the analog signal is converted into a digital signal, the greater the signal-to-noise ratio becomes, as analog signal are easily susceptible to many types of interferences.

An ideal GC detector is distinguished by several characteristics. The first requirement is adequate sensitivity to provide a high resolution signal for all components in the mixture. This is clearly an idealized statement as such a sample would approach zero volume and the detector would need infinite sensitivity to detect it. In modern instruments, the sensitivities of the detectors are in the range of 10^{-8} to 10^{-15} g of solute per second. Furthermore, the quantity of sample must be reproducible and many columns will distort peaks if enough sample is not injected. An ideal column will also be chemically inert and and should not alter the sample in any way. Optimized columns will be able to withstand temperatures in the range of -200 °C to at least 400 °C. In addition, such a column would have a short linear response time that is independent of flow rate and extends for several orders of magnitude. Moreover, the detector should be reliable, predictable and easy to operate.

Understandably, it is not possible for a detector meet all of these requirements.

Table: Typical gas chromatography detectors and their detection limits.

Type of Detector	Applicable Samples	Detection Limit
Mass Spectrometer (MS)	Tunable for any sample	.25 to 100 pg
Flame Ionization (FID)	Hydrocarbons	1 pg/s
Thermal Conductivity (TCD)	Universal	500 pg/ml
Electron-Capture (ECD)	Halogenated hydrocarbons	5 fg/s
Atomic Emission (AED)	Element-selective	1 pg
Chemiluminescence (CS)	Oxidizing reagent	Dark current of PMT
Photoionization (PID)	Vapor and gaseous Compounds	.002 to .02 µg/L

Mass Spectrometry Detectors

Mass Spectrometer (MS) detectors are most powerful of all gas chromatography detectors. In a GC/MS system, the mass spectrometer scans the masses continuously throughout the separation. When the sample exits the chromatography column, it is passed through a transfer line into the inlet of the mass spectrometer . The sample is then ionized and fragmented, typically by an electron-impact ion source. During this process, the sample is bombarded by energetic electrons which ionize the molecule by causing them to lose an electron due to electrostatic repulsion. Further bombardment causes the ions to fragment. The ions are then passed into a mass analyzer where the ions are sorted according to their m/z value, or mass-to-charge ratio. Most ions are only singly charged.

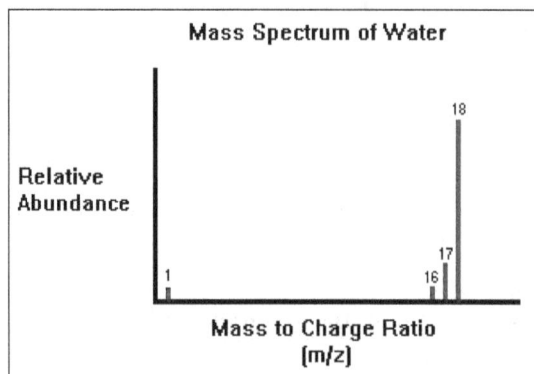

Mass Spectrum of Water.

The Chromatogram will point out the retention times and the mass spectrometer will use the peaks to determine what kind of molecules are exist in the mixture. A typical mass spectrum of water with the absorption peaks at the appropriate m/z ratios.

Instrumentation

One of the most common types of mass analyzer in GC/MS is the quadrupole ion-trap analyzer, which allows gaseous anions or cations to be held for long periods of time by electric and magnetic fields. A simple quadrupole ion-trap consists of a hollow ring electrode with two grounded end-cap electrodes as seen in figure. Ions are allowed into the cavity through a grid in the upper end cap. A variable radio-frequency is applied to the ring electrode and ions with an appropriate m/z value orbit around the cavity. As the radio-frequency is increased linearly, ions of a stable m/z value are ejected by mass-selective ejection in order of mass. Ions that are too heavy or too light are destabilized and their charge is neutralized upon collision with the ring electrode wall. Emitted ions then strike an electron multiplier which converts the detected ions into an electrical signal. This electrical signal is then picked up by the computer through various programs. As an end result, a chromatogram is produced representing the m/z ratio versus the abundance of the sample.

GC/MS units are advantageous because they allow for the immediate determination of the mass of the analyte and can be used to identify the components of incomplete separations. They are rugged, easy to use and can analyze the sample almost as quickly as it is eluted. The disadvantages of mass spectrometry detectors are the tendency for samples to thermally degrade before detection and the end result of obliterating all the sample by fragmentation.

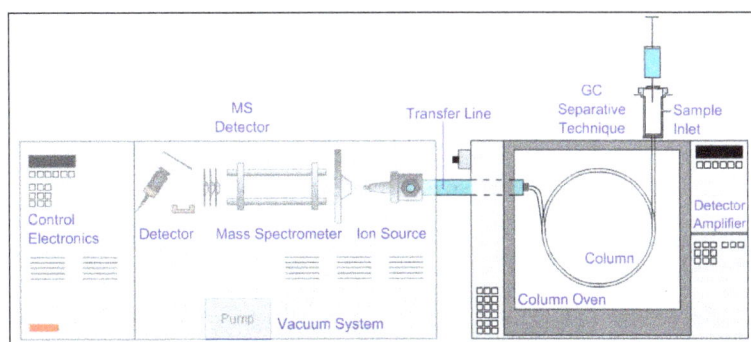

Schematic of the GC/MS system.

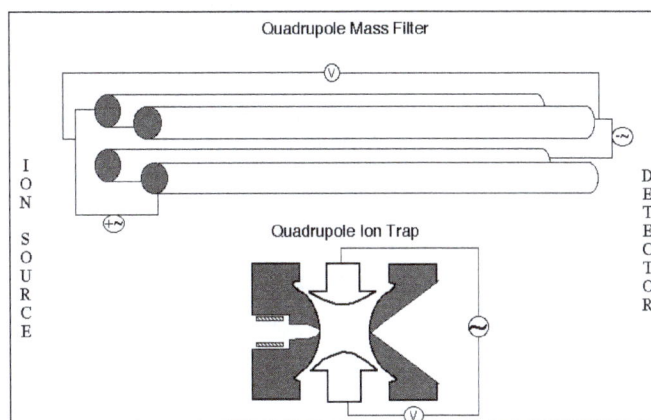

Arrangement of the poles in Quadrupole and Ion Trap Mass spectrometers.

Flame Ionization Detectors

Flame ionization detectors (FID) are the most generally applicable and most widely used detectors. In a FID, the sample is directed at an air-hydrogen flame after exiting the column. At the high temperature of the air-hydrogen flame, the sample undergoes pyrolysis, or chemical decomposition through intense heating. Pyrolized hydrocarbons release ions and electrons that carry current. A high-impedance picoammeter measures this current to monitor the sample's elution.

It is advantageous to used FID because the detector is unaffected by flow rate, noncombustible gases and water. These properties allow FID high sensitivity and low noise. The unit is both reliable and relatively easy to use. However, this technique does require flammable gas and also destroys the sample.

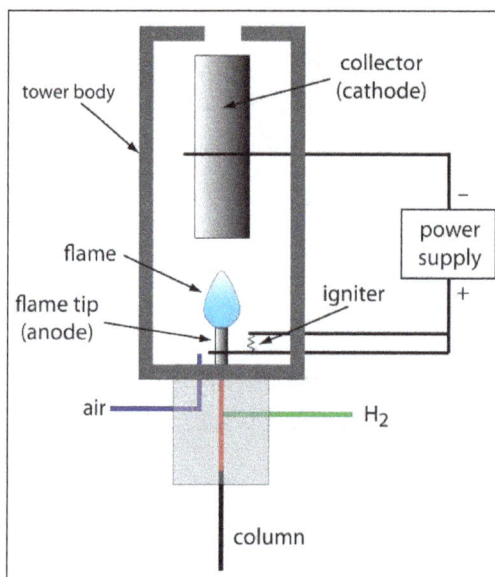

Schematic of a typical flame ionization detector.

Thermal Conductivity Detectors

Thermal conductivity detectors (TCD) were one the earliest detectors developed for use with gas chromatography. The TCD works by measuring the change in carrier gas thermal conductivity caused by the presence of the sample, which has a different thermal conductivity from that of the carrier gas. Their design is relatively simple, and consists of an electrically heated source that is maintained at constant power. The temperature of the source depends upon the thermal conductivities of the surrounding gases. The source is usually a thin wire made of platinum, gold. The resistance within the wire depends upon temperature, which is dependent upon the thermal conductivity of the gas.

TCDs usually employ two detectors, one of which is used as the reference for the carrier gas and the other which monitors the thermal conductivity of the carrier gas and sample mixture. Carrier gases such as helium and hydrogen has very high thermal conductivities so the addition of even a small amount of sample is readily detected.

The advantages of TCDs are the ease and simplicity of use, the devices' broad application to inorganic and organic compounds, and the ability of the analyte to be collected after separation and detection. The greatest drawback of the TCD is the low sensitivity of the instrument in relation to other detection methods, in addition to flow rate and concentration dependency.

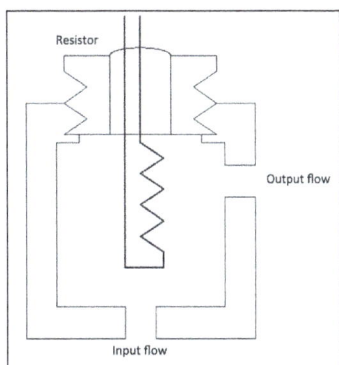

Schematic of thermal conductivity detection cell.

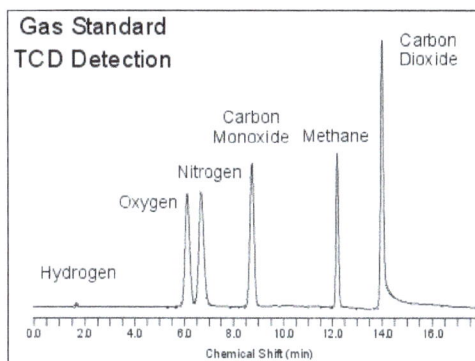

Standard Chromatogram of a Mixture of Gases.

Chromatogram

Figure above represents a standard chromatogram produced by a TCD detector. In a standard chromatogram regardless of the type detector, the x-axis is the time and the y-axis is the abundance or the absorbance. From these chromatograms, retention times and the peak heights are determined and used to further investigate the chemical properties or the abundance of the samples.

Electron-Capture Detectors

Electron-capture detectors (ECD) are highly selective detectors commonly used for detecting environmental samples as the device selectively detects organic compounds with moieties such as halogens, peroxides, quinones and nitro groups and gives little to no response for all other compounds. Therefore, this method is best suited in applications where traces quantities of chemicals such as pesticides are to be detected and other chromatographic methods are unfeasible.

The simplest form of ECD involves gaseous electrons from a radioactive ? emitter in an electric field. As the analyte leaves the GC column, it is passed over this ? emitter, which typically consists of nickle-63 or tritium. The electrons from the ? emitter ionize the nitrogen carrier gas and cause it to release a burst of electrons. In the absence of organic compounds, a constant standing current is maintained between two electrodes. With the addition of organic compounds with electronegative functional groups, the current decreases significantly as the functional groups capture the electrons.

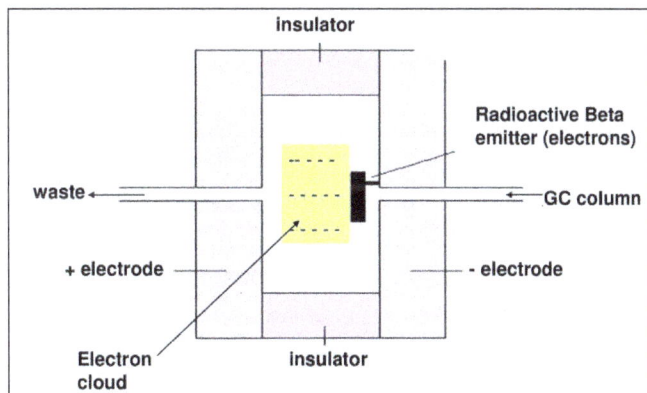

Schematic of an electron-capture detector.

The advantages of ECDs are the high selectivity and sensitivity towards certain organic species with electronegative functional groups. However, the detector has a limited signal range and is potentially dangerous owing to its radioactivity. In addition, the signal-to-noise ratio is limited by radioactive decay and the presence of O_2 within the detector.

Atomic Emission Detectors

Atomic emission detectors (AED), one of the newest addition to the gas chromatographer's arsenal, are element-selective detectors that utilize plasma, which is a partially ionized gas, to atomize all of the elements of a sample and excite their characteristic atomic emission spectra. AED is an extremely powerful alternative that has a wider applicability due to its based on the detection of atomic emissions.There are three ways of generating plasma: microwave-induced plasma (MIP), inductively coupled plasma (ICP) or direct current plasma (DCP). MIP is the most commonly employed form and is used with a positionable diode array to simultaneously monitor the atomic emission spectra of several elements.

Instrumentation

The components of the Atomic emission detectors include 1) an interface for the incoming capillary GC column to induce plasma chamber,2) a microwave chamber, 3) a cooling system, 4) a diffration grating that associated optics, and 5) a position adjustable photodiode array interfaced to a computer.

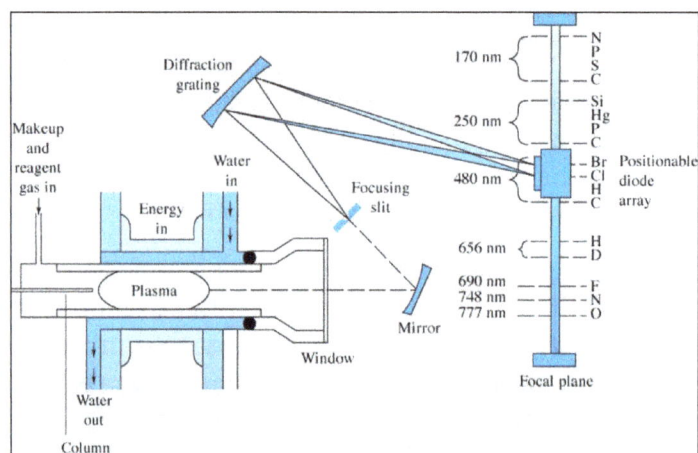

Schematic of atomic emission detector.

GC Chemiluminescence Detectors

Chemiluminescence spectroscopy (CS) is a process in which both qualitative and quantitative properties can be be determined using the optical emission from excited chemical species. It is very similar to AES, but the difference is that it utilizes the light emitted from the energized molecules rather than just excited molecules. Moreover, chemiluminescence can occur in either the solution or gas phase whereas AES is designed for gaseous phases. The light source for chemiluminescence comes from the reactions of the chemicals such that it produces light energy as a product. This light band is used instead of a separate source of light such as a light beam.

Like other methods, CS also has its limitations and the major limitation to the detection limits of CS concerns with the use of a photomultiplier tube (PMT). A PMT requires a dark current in it to detect the light emitted from the analyte.

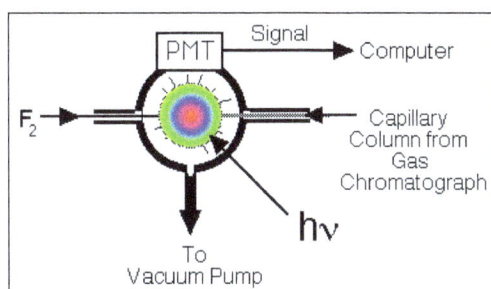

Schematic of a GC Chemiluminescence Detector.

Photoionization Detectors

Another different kind of detector for GC is the photoionization detector which utilizes the properties of chemiluminescence spectroscopy. Photoionization detector (PID) is a portable vapor and gas detector that has selective determination of aromatic hydrocarbons, organo-heteroatom, inorganice species and other organic compounds. PID comprise of an ultrviolet lamp to emit photons that are absorbed by the compounds in an ionization chamber exiting from a GC column. Small fraction of the analyte molecules are actually ionized, nondestructive, allowing confirmation analytical results through other detectors. In addition, PIDs are available in portable hand-held models and in a number of lamp configurations. Results are almost immediate. PID is used commonly to detect VOCs in soil, sediment, air and water, which is often used to detect contaminants in ambient air and soil. The disavantage of PID is unable to detect certain hydrocarbon that has low molecular weight, such as methane and ethane.

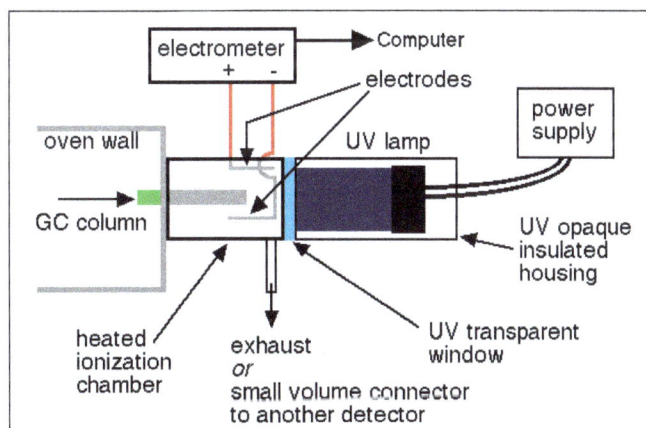

Schematic of a photoionization detector.

Limitations

- Not suitable for detecting semi-volatile compounds.

- Only indicates if volatile organic compounds are presents.

- High concentration so methane are required for higher performance.

- Frequent calibration are required.

- Units of parts per million range.

- Enviromental distraction, especially water vapor.

- Strong electrical fields, Rapid variation in temperature at the detector and naturally occurring compounds may affect instrumental signal.

Height Equivalent to a Theoretical Plate

"Height equivalent to a theoretical plate" (*HETP*) use to calculate the flow rate by using the total number of theoretical plates (*N*) and column length (*L*). Some application, *HETP* concepts is used in industrial practice to convert number of theoretical plates to packing height. *HETP* can be calculate with the Van Deemter equation, which is given by,

$$HETP = A + \frac{B}{v} + Cv$$

Where, *A* and *B* and *C* are constants and v is the linear velocity (carrier flow rate).

- A is the "Eddy-Diffusion" term and causes the broadening of the solute band.

- B is the "Longitudinal diffusion" term whereby the concentration of the analyte, in which diffuses out from the center to the edges.This causes the broadering of the analyte band.

- C is the "Resistance to Mass Transfer" term and causes the band of the analyte broader.

$$HETP = \frac{L}{N}$$

L is the length of the column, where N is the number of theoretical plates, tR is the retention time, and ω is the width of the elution peak at its base.

$$N = 16\left(\frac{tR}{\omega}\right)^2$$

In which, the more plates give a better resolution and more efficiency. Resolution can be determined by,

$$R = 2\left[\frac{(tR)B - (tR)A}{WA + WB}\right]$$

A relationship between the plates and resolution is giving by,

$$R = (N)1/2/4\left(\alpha - \frac{1}{\alpha}\right)\left(1 + \frac{K'B}{K'B}\right)$$

Where the selectivity, a, and k' is the capacity factors take places of the two solutes. The selectivity and capacity factors can be control by improving separation, such as changing mobile/ stationary phase composition, column temperature and use a special chemical effect.

GC Analysis

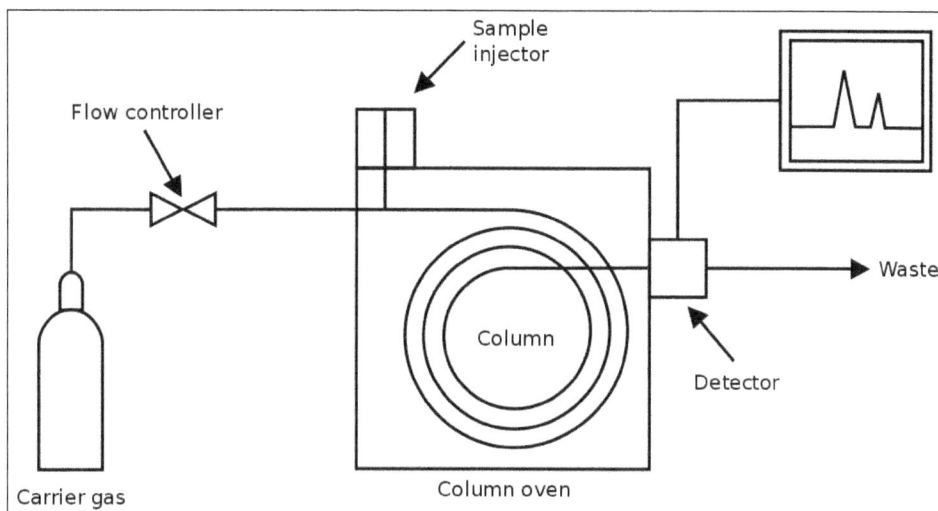

Diagram of a gas chromatograph.

A gas chromatograph is a chemical analysis instrument for separating chemicals in a complex sample. A gas chromatograph uses a flow-through narrow tube known as the column, through which different chemical constituents of a sample pass in a gas stream (carrier gas, mobile phase) at different rates depending on their various chemical and physical properties and their interaction with a specific column filling, called the stationary phase. As the chemicals exit the end of the column, they are detected and identified electronically. The function of the stationary phase in the column is to separate different components, causing each one to exit the column at a different time (retention time). Other parameters that can be used to alter the order or time of retention are the carrier gas flow rate, column length and the temperature.

In a GC analysis, a known volume of gaseous or liquid analyte is injected into the "entrance" (head) of the column, usually using a microsyringe (or, solid phase microextraction fibers, or a gas source switching system). As the carrier gas sweeps the analyte molecules through the column, this motion is inhibited by the adsorption of the analyte molecules either onto the column walls or onto packing materials in the column. The rate at which the molecules progress along the column depends on the strength of adsorption, which in turn depends on the type of molecule and on the stationary phase materials. Since each type of molecule has a different rate of progression, the various components of the analyte mixture are separated as they progress along the column and reach the end of the column at different times (retention time). A detector is used to monitor the outlet stream from the column; thus, the time at which each component reaches the outlet and the amount of that component can be determined. Generally, substances are identified (qualitatively) by the order in which they emerge (elute) from the column and by the retention time of the analyte in the column.

Methods

The method is the collection of conditions in which the GC operates for a given analysis. Method development is the process of determining what conditions are adequate and/or ideal for the analysis required.

This image above shows the interior of a GeoStrata Technologies Eclipse Gas Chromatograph that runs continuously in three-minute cycles. Two valves are used to switch the test gas into the sample loop. After filling the sample loop with test gas, the valves are switched again applying carrier gas pressure to the sample loop and forcing the sample through the column for separation.

Conditions which can be varied to accommodate a required analysis include inlet temperature, detector temperature, column temperature and temperature program, carrier gas and carrier gas flow rates, the column's stationary phase, diameter and length, inlet type and flow rates, sample size and injection technique. Depending on the detector(s) installed on the GC, there may be a number of detector conditions that can also be varied. Some GCs also include valves which can change the route of sample and carrier flow. The timing of the opening and closing of these valves can be important to method development.

Carrier Gas Selection and Flow Rates

Typical carrier gases include helium, nitrogen, argon, hydrogen and air. Which gas to use is usually determined by the detector being used, for example, a DID requires helium as the carrier gas. When analyzing gas samples, however, the carrier is sometimes selected based on the sample's matrix, for example, when analyzing a mixture in argon, an argon carrier is preferred, because the argon in the sample does not show up on the chromatogram. Safety and availability can also influence carrier selection, for example, hydrogen is flammable, and high-purity helium can be difficult to obtain in some areas of the world. As a result of helium becoming more scarce, hydrogen is often being substituted for helium as a carrier gas in several applications.

The purity of the carrier gas is also frequently determined by the detector, though the level of sensitivity needed can also play a significant role. Typically, purities of 99.995% or higher are used. The most common purity grades required by modern instruments for the majority of sensitivities are 5.0 grades, or 99.999% pure meaning that there is a total of 10 ppm of impurities in the carrier gas that could affect the results. The highest purity grades in common use are 6.0 grades, but the need for detection at very low levels in some forensic and environmental applications has driven the need for carrier gases at 7.0 grade purity and these are now commercially available. Trade names for typical purities include "Zero Grade," "Ultra-High Purity (UHP) Grade," "4.5 Grade" and "5.0 Grade."

The carrier gas linear velocity affects the analysis in the same way that temperature does. The higher the linear velocity the faster the analysis, but the lower the separation between analytes. Selecting the linear velocity is therefore the same compromise between the level of separation and length of analysis as selecting the column temperature. The linear velocity will be implemented by means of the carrier gas flow rate, with regards to the inner diameter of the column.

With GCs made before the 1990s, carrier flow rate was controlled indirectly by controlling the carrier inlet pressure, or "column head pressure." The actual flow rate was measured at the outlet of the column or the detector with an electronic flow meter, or a bubble flow meter, and could be an involved, time consuming, and frustrating process. It was not possible to vary the pressure setting during the run, and thus the flow was essentially constant during the analysis. The relation between flow rate and inlet pressure is calculated with Poiseuille's equation for compressible fluids.

Many modern GCs, however, electronically measure the flow rate, and electronically control the carrier gas pressure to set the flow rate. Consequently, carrier pressures and flow rates can be adjusted during the run, creating pressure/flow programs similar to temperature programs.

Stationary Compound Selection

The polarity of the solute is crucial for the choice of stationary compound, which in an optimal case would have a similar polarity as the solute. Common stationary phases in open tubular columns are cyanopropylphenyl dimethyl polysiloxane, carbowax polyethyleneglycol, biscyanopropyl cyanopropylphenyl polysiloxane and diphenyl dimethyl polysiloxane. For packed columns more options are available.

Inlet Types and Flow Rates

The choice of inlet type and injection technique depends on if the sample is in liquid, gas, adsorbed, or solid form, and on whether a solvent matrix is present that has to be vaporized. Dissolved samples can be introduced directly onto the column via a COC injector, if the conditions are well known; if a solvent matrix has to be vaporized and partially removed, a S/SL injector is used (most common injection technique); gaseous samples (e.g., air cylinders) are usually injected using a gas switching valve system; adsorbed samples (e.g., on adsorbent tubes) are introduced using either an external (on-line or off-line) desorption apparatus such as a purge-and-trap system, or are desorbed in the injector (SPME applications).

Sample Size and Injection Technique

Sample Injection

The real chromatographic analysis starts with the introduction of the sample onto the column. The development of capillary gas chromatography resulted in many practical problems with the injection technique. The technique of on-column injection, often used with packed columns, is usually not possible with capillary columns. In the injection system in the capillary gas chromatograph the amount injected should not overload the column and the width of the injected plug should be small compared to the spreading due to the chromatographic process. Failure to comply with this latter requirement will reduce the separation capability of the column. As a general rule, the volume injected, Vinj, and the volume of the detector cell, Vdet, should be about 1/10 of the volume occupied by the portion of sample containing the molecules of interest (analytes) when they exit the column.

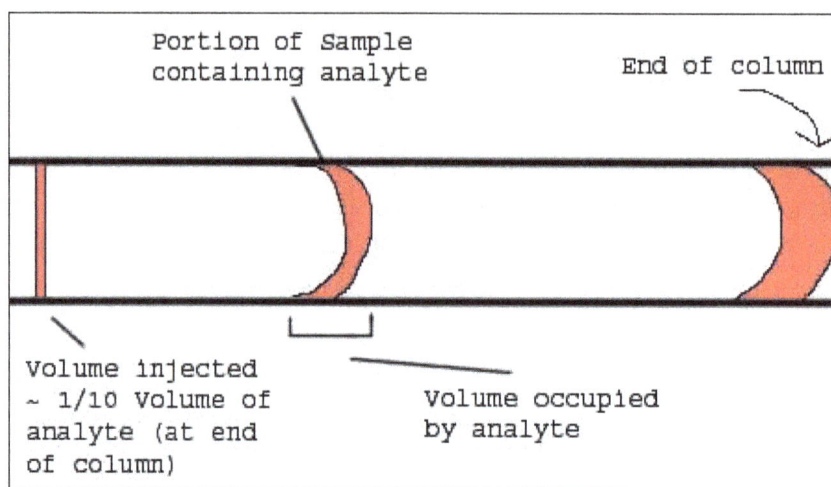

The rule of ten in gas chromatography.

Some general requirements which a good injection technique should fulfill are that it should be possible to obtain the column's optimum separation efficiency, it should allow accurate and reproducible injections of small amounts of representative samples, it should induce no change in sample composition, it should not exhibit discrimination based on differences in boiling point, polarity, concentration or thermal/catalytic stability, and it should be applicable for trace analysis as well as for undiluted samples.

However, there are a number of problems inherent in the use of syringes for injection. Even the best syringes claim an accuracy of only 3%, and in unskilled hands, errors are much larger. The needle may cut small pieces of rubber from the septum as it injects sample through it. These can block the needle and prevent the syringe filling the next time it is used. It may not be obvious that this has happened. A fraction of the sample may get trapped in the rubber, to be released during subsequent injections. This can give rise to ghost peaks in the chromatogram. There may be selective loss of the more volatile components of the sample by evaporation from the tip of the needle.

Column Selection

The choice of column depends on the sample and the active measured. The main chemical attribute regarded when choosing a column is the polarity of the mixture, but functional groups can play a large part in column selection. The polarity of the sample must closely match the polarity of the column stationary phase to increase resolution and separation while reducing run time. The separation and run time also depends on the film thickness (of the stationary phase), the column diameter and the column length.

Column Temperature and Temperature Program

A gas chromatography oven, open to show a capillary column.

The columns in a GC are contained in an oven, the temperature of which is precisely controlled electronically.

The rate at which a sample passes through the column is directly proportional to the temperature of the column. The higher the column temperature, the faster the sample moves through the column. However, the faster a sample moves through the column, the less it interacts with the stationary phase, and the less the analytes are separated.

In general, the column temperature is selected to compromise between the length of the analysis and the level of separation.

A method which holds the column at the same temperature for the entire analysis is called "isothermal." Most methods, however, increase the column temperature during the analysis, the initial temperature, rate of temperature increase (the temperature "ramp"), and final temperature are called the temperature program.

A temperature program allows analytes that elute early in the analysis to separate adequately, while shortening the time it takes for late-eluting analytes to pass through the column.

Data Reduction and Analysis

Qualitative Analysis

Generally, chromatographic data is presented as a graph of detector response (y-axis) against retention time (x-axis), which is called a chromatogram. This provides a spectrum of peaks for a sample representing the analytes present in a sample eluting from the column at different times. Retention time can be used to identify analytes if the method conditions are constant. Also, the pattern of peaks will be constant for a sample under constant conditions and can identify complex mixtures of analytes. However, in most modern applications, the GC is connected to a mass spectrometer or similar detector that is capable of identifying the analytes represented by the peaks.

Quantitative Analysis

The area under a peak is proportional to the amount of analyte present in the chromatogram. By calculating the area of the peak using the mathematical function of integration, the concentration of an analyte in the original sample can be determined. Concentration can be calculated using a calibration curve created by finding the response for a series of concentrations of analyte, or by determining the relative response factor of an analyte. The relative response factor is the expected ratio of an analyte to an internal standard (or external standard) and is calculated by finding the response of a known amount of analyte and a constant amount of internal standard (a chemical added to the sample at a constant concentration, with a distinct retention time to the analyte).

In most modern GC-MS systems, computer software is used to draw and integrate peaks, and match MS spectra to library spectra.

Applications

In general, substances that vaporize below 300 °C (and therefore are stable up to that temperature) can be measured quantitatively. The samples are also required to be salt-free; they should not contain ions. Very minute amounts of a substance can be measured, but it is often required that the sample must be measured in comparison to a sample containing the pure, suspected substance known as a reference standard.

Various temperature programs can be used to make the readings more meaningful; for example to differentiate between substances that behave similarly during the GC process.

Professionals working with GC analyze the content of a chemical product, for example in assuring the quality of products in the chemical industry; or measuring toxic substances in soil, air or water. GC is very accurate if used properly and can measure picomoles of a substance in a 1 ml liquid sample, or parts-per-billion concentrations in gaseous samples.

In practical courses at colleges, students sometimes get acquainted to the GC by studying the contents of Lavender oil or measuring the ethylene that is secreted by *Nicotiana benthamiana* plants after artificially injuring their leaves. These GC analyse hydrocarbons (C_2-C_{40}+). In a typical experiment, a packed column is used to separate the light gases, which are then detected with a TCD. The hydrocarbons are separated using a capillary column and detected with a FID. A complication with light gas analyses that include H_2 is that He, which is the most common and most sensitive inert carrier (sensitivity is proportional to molecular mass) has an almost identical thermal conductivity to hydrogen (it is the difference in thermal conductivity between two separate filaments in a Wheatstone Bridge type arrangement that shows when a component has been eluted). For this reason, dual TCD instruments used with a separate channel for hydrogen that uses nitrogen as a carrier are common. Argon is often used when analysing gas phase chemistry reactions such as F-T synthesis so that a single carrier gas can be used rather than two separate ones. The sensitivity is reduced, but this is a trade off for simplicity in the gas supply.

Gas Chromatography is used extensively in forensic science. Disciplines as diverse as solid drug dose (pre-consumption form) identification and quantification, arson investigation, paint chip analysis, and toxicology cases, employ GC to identify and quantify various biological specimens and crime-scene evidence.

GC Detectors

As solutes elute from the column, they interact with the detector. The detector converts this interaction into an electronic signal that is sent to the data system. The magnitude of the signal is plotted versus time (from the time of injection) and a chromatogram is generated.

Some detectors respond to any solute eluting from the column while others respond only to solutes with specific structures, functional groups or atoms.

Detectors that exhibit enhanced response to specific types of solutes are called selective detectors.

Most detectors require one or more gases to function properly. There are combustion, reagent, auxiliary and makeup gases. In some cases, one gas may serve multiple purposes. The type of detector gas is dependent on the specific detector and is fairly universal between GC manufacturers. The flow rates for each type of detector varies between GC manufacturers. It is important to follow the recommended flow rates to obtain the optimal sensitivity, selectivity and linear range for a detector.

Flame Ionization Detector (FID)

Mechanism: Compounds are burned in a hydrogen-air flame. Carbon containing compounds produce ions that are attracted to the collector. The number of ions hitting the collector is measured and a signal is generated.

Selectivity: Compounds with C-H bonds. A poor response for some non-hydrogen containing organics (e.g., hexachlorobenzene).

- Sensitivity: 0.1-10 ng.

- Linear range: 105-107.

- Gases: Combustion - hydrogen and air; Makeup - helium or nitrogen.

- Temperature: 250-300 °C,and 400-450 °C for high temperature analyses.

Nitrogen Phosphorus Detector (NPD)

Mechanism: Compounds are burned in a plasma surrounding a rubidium bead supplied with hydrogen and air. Nitrogen and phosphorous containing compounds produce ions that are attracted to the collector. The number of ions hitting the collector is measured and a signal is generated.

- Selectivity: Nitrogen and phosphorous containing compounds.

- Sensitivity: 1-10 pg.

- Linear range: 10^4-10^{-6}.

- Gases: Combustion - hydrogen and air; Makeup - helium.

- Temperature: 250-300 °C.

Electron Capture Detector (ECD)

Mechanism: Electrons are supplied from a 63Ni foil lining the detector cell. A current is generated in the cell. Electronegative compounds capture electrons resulting in a reduction in the current. The amount of current loss is indirectly measured and a signal is generated.

- Selectivity: Halogens, nitrates and conjugated carbonyls.

- Sensitivity: 0.1-10 pg (halogenated compounds); 1-100 pg.

- (nitrates): 0.1-1 ng (carbonyls).

- Linear range: 10^3-10^4.

- Gases: Nitrogen or argon/methane.

- Temperature: 300-400 °C.

Thermal Conductivity Detector (TCD)

Mechanism: A detector cell contains a heated filament with an applied current. As carrier gas containing solutes passes through the cell, a change in the filament current occurs. The current change is compared against the current in a reference cell. The difference is measured and a signal is generated.

- Selectivity: All compounds except for the carrier gas.

- Sensitivity: 5-20 ng.

- Linear range: 10^5-10^6 .

- Gases: Makeup - same as the carrier gas.

- Temperature: 150-250 °C.

Flame Photometric Detector (FPD)

Mechanism: Compounds are burned in a hydrogen-air flame. Sulfur and phosphorous containing compounds produce light emitting species (sulfur at 394 nm and phosphorous at 526 nm). A monochromatic filter allows only one of the wavelengths to pass. A photomultiplier tube is used to measure the amount of light and a signal is generated. A different filter is required for each detection mode.

- Selectivity: Sulfur or phosphorous containing compounds. Only one at a time.

- Sensitivity: 10-100 pg (sulfur); 1-10 pg (phosphorous).

- Linear range: Non-linear (sulfur); 10^3-10^5 (phosphorous).

- Gases: Combustion - hydrogen and air; Makeup - nitrogen.

- Temperature: 250-300 °C.

Photoionization Detector (PID)

Mechanism: Compounds eluting into a cell are bombarded with high energy photons emitted from a lamp. Compounds with ionization potentials below the photon energy are ionized. The resulting ions are attracted to an electrode, measured, and a signal is generated.

- Selectivity: Depends on lamp energy. Usually used for aromatics and olefins (10 eV lamp).

- Sensitivity: 25-50 pg (aromatics); 50-200 pg (olefins).

- Linear range: 10^5-10^6.

- Gases: Makeup - same as the carrier gas.

- Temperature: 200 °C.

Electrolytic Conductivity Detector (ELCD)

Mechanism: Compounds are mixed with a reaction gas and passed through a high temperature reaction tube. Specific reaction products are created which mix with a solvent and pass through an electrolytic conductivity cell. The change in the electrolytic conductivity of the solvent is measured and a signal is generated. Reaction tube temperature and solvent determine which types of compounds are detected.

- Selectivity: Halogens, sulfur or nitrogen containing compounds. Only one at a time.

- Sensitivity: 5-10 pg (halogens); 10-20 pg (S); 10-20 pg (N).

- Linear range: 10^5-10^6 (halogens); 10^4-10^5 (N); $10^{3.5}$-10^4(S).

- Gases: Hydrogen (halogens and nitrogen); air (sulfur).

- Temperature: 800-1000 °C (halogens), 850-925 °C (N), 750-825 °C (S).

Mass Spectrometer (MS)

Mechanism: The detector is maintained under vacuum. Compounds are bombarded with electrons (EI) or gas molecules (CI). Compounds fragment into characteristic charged ions or fragments. The resulting ions are focused and accelerated into a mass filter. The mass filter selectively allows all ions of a specific mass to pass through to the electron multiplier. All of the ions of the specific mass are detected. The mass filter then allows the next mass to pass through while excluding all others. The mass filter scans stepwise through the designated range of masses several times per second. The total number of ions are counted for each scan. The abundance or number of ions per scan is plotted versus time to obtain the chromatogram (called the TIC). A mass spectrum is obtained for each scan which plots the various ion masses versus their abundance or number.

- Selectivity: Any compound that produces fragments within the selected mass range. May be an inclusive range of masses (full scan) or only select ions (SIM).

- Sensitivity: 1-10 ng (full scan); 1-10 pg (SIM).

- Linear range: 10^5-10^6.

- Gases: None.

- Temperature: 250-300 °C (transfer line), 150-250 °C.

Principle of Gas Chromatography

The equilibrium for gas chromatography is partitioning, and the components of the sample will partition (i.e. distribute) between the two phases: the stationary phase and the mobile phase.

Compounds that have greater affinity for the stationary phase spend more time in the column and thus elute later and have a longer retention time (tR) than samples that have higher affinity for the mobile phase.

Affinity for the stationary phase is driven mainly by intermolecular interactions and the polarity of the stationary phase can be chosen to maximize interactions and thus the separation.

Physical Components of GC

Autosamplers

The autosampler provides the means to introduce a sample automatically into the inlets. Manual insertion of the sample is possible but is no longer common. Automatic insertion provides better reproducibility and time-optimization.

Different kinds of autosamplers exist. Autosamplers can be classified in relation to sample capacity (auto-injectors vs. autosamplers, where auto-injectors can work a small number of samples), to robotic technologies (XYZ robot vs. rotating robot – the most common), or to analysis:

- Liquid.

- Static head-space by syringe technology.

- Dynamic head-space by transfer-line technology.

- Solid phase microextraction (SPME).

Inlets

The column inlet (or injector) provides the means to introduce a sample into a continuous flow of carrier gas. The inlet is a piece of hardware attached to the column head.

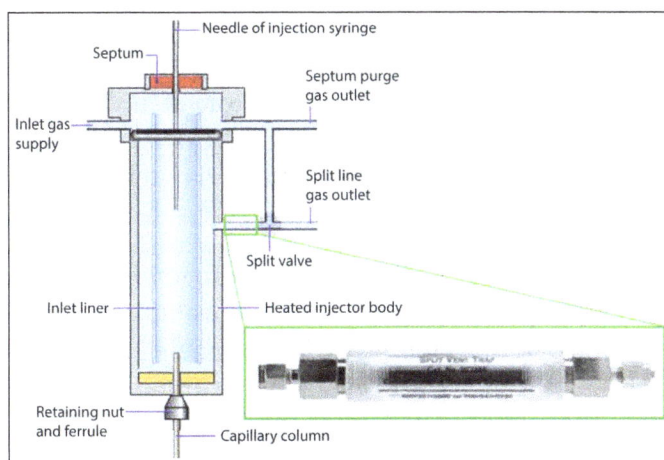

Split/splitless inlet.

Common inlet types are:

- S/SL (split/splitless) injector: A sample is introduced into a heated small chamber via a syringe through a septum – the heat facilitates volatilization of the sample and sample matrix. The carrier gas then either sweeps the entirety (splitless mode) or a portion (split mode) of the sample into the column. In split mode, a part of the sample/carrier gas mixture in the injection chamber is exhausted through the split vent. Split injection is preferred when working with samples with high analyte concentrations (>0.1%) whereas splitless injection is best suited for trace analysis with low amounts of analytes (<0.01%). In splitless mode the split valve opens after a pre-set amount of time to purge heavier elements that would otherwise contaminate the system. This pre-set (splitless) time should be optimized, the shorter time (e.g., 0.2 min) ensures less tailing but loss in response, the longer time (2 min) increases tailing but also signal.

- On-column inlet: The sample is here introduced directly into the column in its entirety without heat, or at a temperature below the boiling point of the solvent. The low temperature condenses the sample into a narrow zone. The column and inlet can then be heated, releasing the sample into the gas phase. This ensures the lowest possible temperature for chromatography and keeps samples from decomposing above their boiling point.

- PTV injector: Temperature-programmed sample introduction was first described by Vogt in 1979. Originally Vogt developed the technique as a method for the introduction of large sample volumes (up to 250 µL) in capillary GC. Vogt introduced the sample into the liner at a controlled injection rate. The temperature of the liner was chosen slightly below the boiling point of the solvent. The low-boiling solvent was continuously evaporated and vented through the split line. Based on this technique, Poy developed the programmed temperature vaporising injector; PTV. By introducing the sample at a low initial liner temperature many of the disadvantages of the classic hot injection techniques could be circumvented.

- Gas source inlet or gas switching valve: Gaseous samples in collection bottles are connected to what is most commonly a six-port switching valve. The carrier gas flow is not interrupted while a sample can be expanded into a previously evacuated sample loop. Upon switching, the contents of the sample loop are inserted into the carrier gas stream.

- P/T (Purge-and-Trap) system: An inert gas is bubbled through an aqueous sample causing insoluble volatile chemicals to be purged from the matrix. The volatiles are 'trapped' on an absorbent column (known as a trap or concentrator) at ambient temperature. The trap is then heated and the volatiles are directed into the carrier gas stream. Samples requiring preconcentration or purification can be introduced via such a system, usually hooked up to the S/SL port.

The choice of carrier gas (mobile phase) is important. Hydrogen has a range of flow rates that are comparable to helium in efficiency. However, helium may be more efficient and provide the best separation if flow rates are optimized. Helium is non-flammable and works with a greater number of detectors and older instruments. Therefore, helium is the most common carrier gas used. However, the price of helium has gone up considerably over recent years, causing an increasing number of chromatographers to switch to hydrogen gas. Historical use, rather than rational consideration, may contribute to the continued preferential use of helium.

Detectors

The most commonly used detectors are the flame ionization detector (FID) and the thermal conductivity detector (TCD). Both are sensitive to a wide range of components, and both work over a wide range of concentrations. While TCDs are essentially universal and can be used to detect any component other than the carrier gas (as long as their thermal conductivities are different from that of the carrier gas, at detector temperature), FIDs are sensitive primarily to hydrocarbons, and are more sensitive to them than TCD. However, a FID cannot detect water. Both detectors are also quite robust. Since TCD is non-destructive, it can be operated in-series before a FID (destructive), thus providing complementary detection of the same analytes. Other detectors are sensitive only to specific types of substances, or work well only in narrower ranges of concentrations.

Thermal conductivity detector (TCD) relies on the thermal conductivity of matter passing around a tungsten -rhenium filament with a current traveling through it. In this set up helium or nitrogen serve as the carrier gas because of their relatively high thermal conductivity which keep the filament cool and maintain uniform resistivity and electrical efficiency of the filament. However, when analyte molecules elute from the column, mixed with carrier gas, the thermal conductivity decreases and this causes a detector response. The response is due to the decreased thermal conductivity causing an increase in filament temperature and resistivity resulting in fluctuations in voltage. Detector sensitivity is proportional to filament current while it is inversely proportional to the immediate environmental temperature of that detector as well as flow rate of the carrier gas.

In a flame ionization detector (FID), electrodes are placed adjacent to a flame fueled by hydrogen/ air near the exit of the column, and when carbon containing compounds exit the column they are pyrolyzed by the flame. This detector works only for organic/hydrocarbon containing compounds due to the ability of the carbons to form cations and electrons upon pyrolysis which generates a current between the electrodes. The increase in current is translated and appears as a peak in a chromatogram. FIDs have low detection limits (a few picograms per second) but they are unable to generate ions from carbonyl containing carbons. FID compatible carrier gasses include helium, hydrogen, nitrogen, and argon.

Alkali flame detector (AFD) or alkali flame ionization detector (AFID) has high sensitivity to nitrogen and phosphorus, similar to NPD. However, the alkaline metal ions are supplied with the hydrogen gas, rather than a bead above the flame. For this reason AFD does not suffer the "fatigue" of the NPD, but provides a constant sensitivity over long period of time. In addition, when alkali ions are not added to the flame, AFD operates like a standard FID. A catalytic combustion detector (CCD) measures combustible hydrocarbons and hydrogen. Discharge ionization detector (DID) uses a high-voltage electric discharge to produce ions.

The polyarc reactor is an add-on to new or existing GC-FID instruments that converts all organic compounds to methane molecules prior to their detection by the FID. This technique can be used to improve the response of the FID and allow for the detection of many more carbon-containing compounds. The complete conversion of compounds to methane and the now equivalent response in the detector also eliminates the need for calibrations and standards because response factors are all equivalent to those of methane. This allows for the rapid analysis of complex mixtures that contain molecules where standards are not available.

Flame photometric detector (FPD) uses a photomultiplier tube to detect spectral lines of the compounds as they are burned in a flame. Compounds eluting off the column are carried into a hydrogen fueled flame which excites specific elements in the molecules, and the excited elements (P,S, Halogens, Some Metals) emit light of specific characteristic wavelengths. The emitted light is filtered and detected by a photomultiplier tube. In particular, phosphorus emission is around 510–536 nm and sulfur emission is at 394 nm. With an atomic emission detector (AED), a sample eluting from a column enters a chamber which is energized by microwaves that induce a plasma. The plasma causes the analyte sample to decompose and certain elements generate an atomic emission spectra. The atomic emission spectra is diffracted by a diffraction grating and detected by a series of photomultiplier tubes or photo diodes.

Electron capture detector (ECD) uses a radioactive beta particle (electron) source to measure the degree of electron capture. ECD are used for the detection of molecules containing electronegative/withdrawing elements and functional groups like halogens, carbonyl, nitriles, nitro groups, and organometalics. In this type of detector either nitrogen or 5% methane in argon is used as the mobile phase carrier gas. The carrier gas passes between two electrodes placed at the end of the column, and adjacent to the anode (negative electrode) resides a radioactive foil such as 63Ni. The radioactive foil emits a beta particle (electron) which collides with and ionizes the carrier gas to generrate more ions resulting in a current. When analyte molecules with electronegative/withdrawing elements or functional groups electrons are captured which results in a decrease in current generating a detector response.

Nitrogen–phosphorus detector (NPD), a form of thermionic detector where nitrogen and phosphorus alter the work function on a specially coated bead and a resulting current is measured.

Dry electrolytic conductivity detector (DELCD) uses an air phase and high temperature (v. Coulsen) to measure chlorinated compounds.

Mass spectrometer (MS), also called GC-MS: Highly effective and sensitive, even in a small quantity of sample. This detector can be used to identify the analytes in chromatograms by their mass spectrum. Some GC-MS are connected to an NMR spectrometer which acts as a backup detector. This combination is known as GC-MS-NMR. Some GC-MS-NMR are connected to an infrared spectro-photometer which acts as a backup detector. This combination is known as GC-MS-NMR-IR. It must, however, be stressed this is very rare as most analyses needed can be concluded via purely GC-MS.

Vacuum ultraviolet (VUV) represents the most recent development in gas chromatography detectors. Most chemical species absorb and have unique gas phase absorption cross sections in the approximately 120–240 nm VUV wavelength range monitored. Where absorption cross sections are known for analytes, the VUV detector is capable of absolute determination (without calibration) of the number of molecules present in the flow cell in the absence of chemical interferences.

Other detectors include the Hall electrolytic conductivity detector (ElCD), helium ionization detector (HID), infrared detector (IRD), photo-ionization detector (PID), pulsed discharge ionization detector (PDD), and thermionic ionization detector (TID).

Instrumental Components

Carrier Gas

The carrier gas must be chemically inert. Commonly used gases include nitrogen, helium, argon, and carbon dioxide. The choice of carrier gas is often dependant upon the type of detector which is used. The carrier gas system also contains a molecular sieve to remove water and other impurities.

Sample Injection Port

For optimum column efficiency, the sample should not be too large, and should be introduced onto the column as a "plug" of vapour - slow injection of large samples causes band broadening and loss of resolution. The most common injection method is where a microsyringe is used to inject sample through a rubber septum into a flash vapouriser port at the head of the column. The temperature of the sample port is usually about 50 °C higher than the boiling point of the least volatile component of the sample. For packed columns, sample size ranges from tenths of a microliter up to 20 microliters. Capillary columns, on the other hand, need much less sample, typically around 10-3 mL. For capillary GC, split/splitless injection is used.

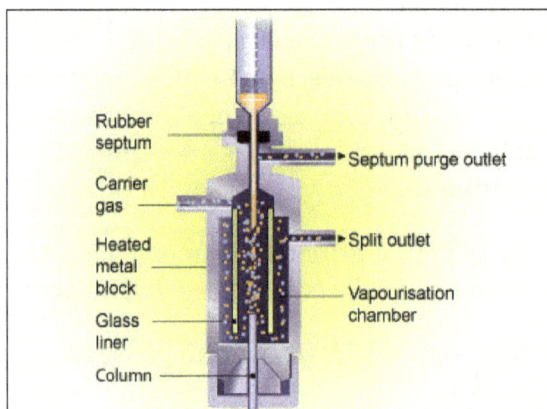

The injector can be used in one of two modes; split or splitless. The injector contains a heated chamber containing a glass liner into which the sample is injected through the septum. The carrier gas enters the chamber and can leave by three routes (when the injector is in split mode). The sample vapourises to form a mixture of carrier gas, vapourised solvent and vapourised solutes. A proportion of this mixture passes onto the column, but most exits through the split outlet. The septum purge outlet prevents septum bleed components from entering the column.

Columns

There are two general types of column, packed and capillary (also known as open tubular). Packed columns contain a finely divided, inert, solid support material (commonly based on diatomaceous earth) coated with liquid stationary phase. Most packed columns are 1.5 - 10 m in length and have an internal diameter of 2 - 4 mm.

Capillary columns have an internal diameter of a few tenths of a millimeter. They can be one of two types; wall-coated open tubular (WCOT) or support-coated open tubular (SCOT). Wall-coated columns consist of a capillary tube whose walls are coated with liquid stationary phase. In support-coated columns, the inner wall of the capillary is lined with a thin layer of support material such as diatomaceous earth, onto which the stationary phase has been adsorbed. SCOT columns are generally less efficient than WCOT columns. Both types of capillary column are more efficient than packed columns.

In 1979, a new type of WCOT column was devised - the Fused Silica Open Tubular (FSOT) column.

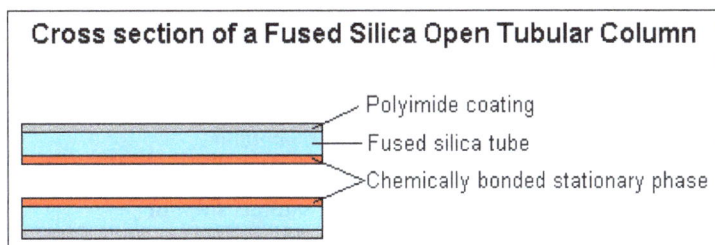

These have much thinner walls than the glass capillary columns, and are given strength by the polyimide coating. These columns are flexible and can be wound into coils. They have the advantages of physical strength, flexibility and low reactivity.

Column Temperature

For precise work, column temperature must be controlled to within tenths of a degree. The optimum column temperature is dependant upon the boiling point of the sample. As a rule of thumb, a temperature slightly above the average boiling point of the sample results in an elution time of 2 - 30 minutes. Minimal temperatures give good resolution, but increase elution times. If a sample has a wide boiling range, then temperature programming can be useful. The column temperature is increased (either continuously or in steps) as separation proceeds.

Detectors

There are many detectors which can be used in gas chromatography. Different detectors will give different types of selectivity. A non-selective detector responds to all compounds except the carrier gas, a selective detector responds to a range of compounds with a common physical or chemical property and a specific detector responds to a single chemical compound. Detectors can also be grouped into concentration dependant detectors and mass flow dependant detectors. The signal from a concentration dependant detector is related to the concentration of solute in the detector, and does not usually destroy the sample Dilution of with make-up gas will lower the detectors response. Mass flow dependant detectors usually destroy the sample, and the signal is related to the rate at which solute molecules enter the detector. The response of a mass flow dependant detector is unaffected by make-up gas. Have a look at this tabular summary of common GC detectors:

Detector	Type	Support gases	Selectivity	Detectability	Dynamic range
Flame ionization (FID)	Mass flow	Hydrogen and air	Most organic cpds.	100 pg	10^7
Thermal conductivity (TCD)	Concentration	Reference	Universal	1 ng	10^7
Electron capture (ECD)	Concentration	Make-up	Halides, nitrates, nitriles, peroxides, anhydrides, organometallics	50 fg	10^5
Nitrogen-phosphorus	Mass flow	Hydrogen and air	Nitrogen, phosphorus	10 pg	10^6
Flame photometric (FPD)	Mass flow	Hydrogen and air possibly oxygen	Sulphur, phosphorus, tin, boron, arsenic, germanium, selenium, chromium	100 pg	10^3
Photo-ionization (PID)	Concentration	Make-up	Aliphatics, aromatics, ketones, esters, aldehydes, amines, heterocyclics, organosulphurs, some organometallics	2 pg	10^7
Hall electrolytic conductivity	Mass flow	Hydrogen, oxygen	Halide, nitrogen, nitrosamine, sulphur		

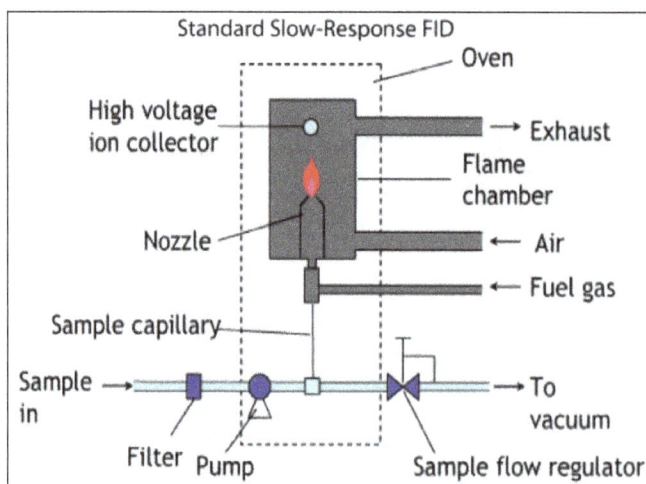

The flame ionization detector.

The effluent from the column is mixed with hydrogen and air, and ignited. Organic compounds burning in the flame produce ions and electrons which can conduct electricity through the flame. A large electrical potential is applied at the burner tip, and a collector electrode is located above the flame. The current resulting from the pyrolysis of any organic compounds is measured. FIDs are mass sensitive rather than concentration sensitive; this gives the advantage that changes in mobile phase flow rate do not affect the detector's response. The FID is a useful general detector for the analysis of organic compounds; it has high sensitivity, a large linear response range, and low noise. It is also robust and easy to use, but unfortunately, it destroys the sample.

Kovats Retention Index

In gas chromatography, Kovats retention index (shorter Kovats index, retention index; plural retention indices) is used to convert retention times into system-independent constants. The index is named after the Hungarian-born Swiss chemist Ervin Kováts, who outlined this concept during the 1950s while performing research into the composition of the essential oils.

The retention index of a certain chemical compound is its retention time normalised to the retention times of adjacently eluting n-alkanes. While retention times vary with the individual chromatographic system (e.g. with regards to column length, film thickness, diameter, carrier gas velocity and pressure, and void time), the derived retention indices are quite independent of these parameters and allow comparing values measured by different analytical laboratories under varying conditions and analysis times from seconds to hours. Tables of retention indices are used to identify peaks by comparing measured retention indices with the tabulated values.

Expression

The Kovats index applies to organic compounds. The method interpolates peaks between bracketing n-alkanes. Kovats index of n-alkanes is carbon number times 100 like n-Butane index is 400. The Kovats index is dimensionless, contrary to retention time or retention volume. For isothermal gas chromatography, the Kovats index is given by the equation:

$$I_i = 100\left[n + (N-n)\frac{log(t_i - t_0) - log(t_n - t_0)}{log(t_N - t_0) - log(t_n - t_0)}\right]$$

Symbols stand for:

I_i the Kováts retention index of peak i;

n carbon number of n-alkane peak heading peak i;

N carbon number of n-alkane peak trailing peak i;

t_i retention time of compound i, minutes;

t_0 air peak, void time in average velocity $u = L/t_0$, minutes.

Kovats and Properties

Compounds elute in the carrier gas phase only. Compounds solved in the stationary phase stay put. The ratio of gas time t_0 and residence time $t_i - t_0$ in the stationary liquid polymer phase is called the capacity factor k_i:

$$k_i = \frac{t_i - t_0}{t_0} = \frac{RTS_i}{P^i}\beta$$

Symbols represent physical properties:

- R gas constant (8.314 J/mole/k);

- T temperature [k];

- S_i solubility of compound i in polymer stationary phase [mole/m³];

- P^i vapor pressure of pure liquid i [Pa].

Capillary tubes with uniform coatings have this phase ratio ß:

$$\beta = \frac{V_L}{V_G} = \frac{4d_f}{d_c}$$

Capillary inner diameter d_c is well defined but film thickness d_c reduces by bleed and thermal breakdown that occur after column heating over time, depending on chemical bonding to the silica glass wall and polymer cross-linking of the stationary phase. Above capacity factor d_c can be expressed explicit for retention time:

$$t_i = t_0 (\frac{RTS_i}{P^i} \frac{4d_f}{d_c} + 1)$$

Retention time t_i is shorter by reduced d_f over column life time. Column length L is introduced with average gas velocity $u = L / t_0$:

$$t_i = \frac{L}{u}(\frac{RTS_i}{P^i} \frac{4d_f}{d_c} + 1)$$

and temperature T have a direct relation with t_i. However, warmer columns $T \uparrow$ do not have longer t_i but shorter, following temperature programming experience. Pure liquid vapor pressure.

P^i rises exponentially with T so that we do get shorter t_i warming the column $T \uparrow$ Solubility of compounds S_i in the stationary phase may rise or fall with T, but not exponentially. S_i is referred to as selectivity or polarity by gas chromatographers today. Isothermal Kovats index in terms of the physical properties becomes:

$$I_i = 100 \left[n + \frac{log(S_i / P^i) - log(S_n / P^n)}{log(S_{n+1} / P^{n+1}) - log(S_n / P^n)} \right]$$

Isothermal Kovats index is independent of R, any GC dimension L or ß or carrier gas velocity u, which compares favorable to retention time t_i. Isothermal Kovats index is based on solubility S_i and vapor pressure P^i of compound i and n-Alkanes $(i = n)$. T dependence depends on the compound compared to the n-alkanes. Kovats index of n-alkanes $I_n = 100c$ is independent of T. Isothermal Kovats indices of hydrocarbon were measured by Axel Lubeck and Donald Sutton.

Temperature Programmed Kovats Index

ASTM method D 6730 defines the temperature programmed chromatography Kovats index equation:

$$I_i = 100 \left[n + \frac{log(t_i) - log(t_n)}{log(t_{n+1}) - log(t_n)} \right]$$

t_n & t_{n+1} retention times of heading and trailing n-alkanes.

TPGC index does depend on temperature program, gas velocity and the column used.

Method Translation Proper method translation aims at keeping the same retention temperatures in a smaller and faster column. Downsize the column of your certified method using method translation to ensure that the on-line fast analysis index corresponds to the certified lab method index. Method translation rules are incorporated in some chromatography data systems.

Gas Chromatography–Vacuum Ultraviolet Spectroscopy

Gas Chromatography–Vacuum Ultraviolet Spectroscopy (GC-VUV) is a universal detection platform for gas chromatography. VUV detection provides both qualitative and quantitative spectral information for most gas phase compounds.

GC-VUV spectral data is three-dimensional (time, absorbance, wavelength) and specific to chemical structure. Nearly all compounds absorb in the VUV region of the electromagnetic spectrum with the exception of carrier gases hydrogen, helium, and argon. The high energy, short wavelength VUV photons probe electronic transitions in almost all chemical bonds including ground state to excited state. The result is spectral "fingerprints" that are specific to individual compound structure and can be readily identified by the VUV library.

Unique VUV spectra enable closely related compounds such as structural isomers to be clearly differentiated. VUV detectors complement mass spectrometry, which struggles with characterizing constitutional isomers and compounds with low mass quantitation ions. VUV spectra can also be used to deconvolve analyte co-elution, resulting in an accurate quantitative representation of individual analyte contribution to the original response. This characteristically lends itself to significantly reducing GC runtimes through flow rate-enhanced chromatographic compression.

VUV spectroscopy follows the simple linear relationship between absorbance and concentration described by the Beer-Lambert Law, resulting in more accurate retention time-based identification. VUV absorbance spectra also exhibit feature similarity within compound classes, meaning VUV detectors can rapidly compound class characterization in complex samples through compound spectral shape and retention index information. Advances in technology reduces the typical group analysis data processing time from 15-30 minutes to <1 minute per sample.

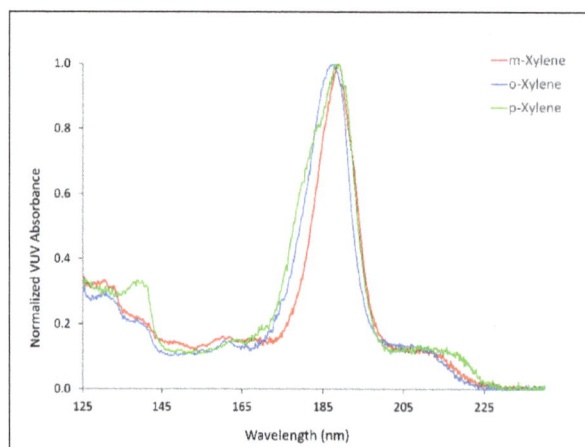

The distinct VUV spectra of m-, p-, and o-xylene. The compounds differ by only the positions of two methyl groups on a benzene ring.

VUV Spectral Identification

Gas phase species absorb and display unique spectra between 120 – 240 nm where high energy $\sigma \rightarrow \sigma^*$, $n \rightarrow \sigma^*$, $\pi \rightarrow \pi^*$, $n \rightarrow \pi^*$ electronic transitions can be excited and probed. VUV spectra reflect the absorbance cross section of compounds and are specific to their electronic structure and functional group arrangement. The ability of VUV detectors to produce spectra for most compounds results in universal and highly selective compound identification. VUV spectroscopy data is highly characteristic while also providing quantitative information. Many commonly used GC detectors such as the electron capture detector (ECD), flame ionization detector (FID), and thermal conductivity detector (TCD) produce quantitative but not qualitative detail. Gas chromatography–mass spectrometry (GC-MS) generates qualitative and quantitative data but has difficulty characterizing labile and low mass compounds, as well as differentiating between isomers. GC-VUV complements MS by overcoming its limitations and providing a secondary method of confirmation. It also offers a single instrument alternative to the use of multiple detectors for qualitative and quantitative analysis.

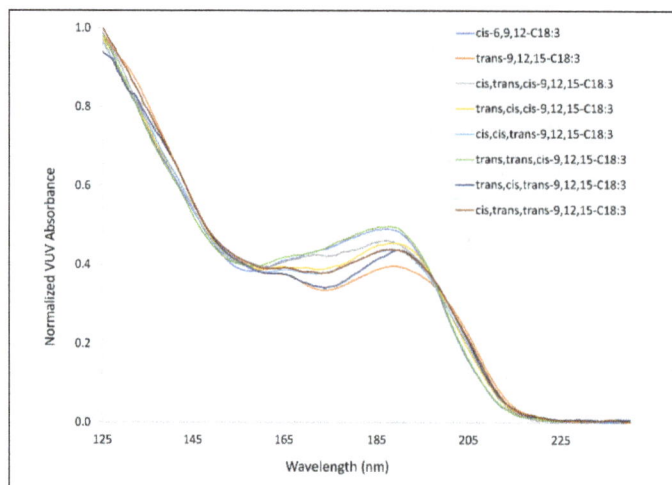

VUV spectra of fatty acid methyl ester (FAME) cis and trans isomers that are commonly found in butter and vegetable oils. GC-VUV can readily distinguish between the C18:3 FAME isomers, cis and trans classification, and the degree of unsaturation.

Naphthols, xylenes, and cis- and trans- fatty acids are compounds that are prohibitively difficult to distinguish according to their electron ionization mass spectral profiles. Xylenes present the additional challenge of natural co-elution that makes separating their isoforms problematic. Figure shows the distinct VUV spectra of m-, p-, and o-xylene. These compounds can be differentiated despite their only difference being the position of two methyl groups around a benzene ring. The spectral differences of these isomers enable their co-elution to be resolved through spectral deconvolution.

Fatty acid screening and profiling is an application that commonly requires the use of multiple detectors to achieve quantitative and qualitative results. FID is a quantitative detector that is suitable for routine screening when guided by retention index information. GC-MS has traditionally been used for qualitative compound profiling, but falls short where isobaric analytes are prevalent. It especially struggles with differentiating cis and trans fatty acid isomers. Electron impact ionization can also cause double bond migration and lead to ambiguous fatty acid structural data.

Determining cis and trans fatty acid distribution in oils and fats is important in assessing their potential health impacts. VUV spectra of trans-containing fatty acid methyl ester (FAME) isomers typically found in butter and vegetable oils are shown in figure. These trans-containing isomers separate chromatographically from cis-containing isomers and have the tendency to co-elute with each other and, in some cases, with select C20:1 isomers. GC-VUV is not only able to differentiate the C18:3 FAME variants, but is also capable of telling cis isomers apart from trans isomers. Degrees of unsaturation such as C20:1 vs. C18:3 can additionally be distinguished. Previous work has demonstrated how distinct VUV spectra enable straightforward deconvolution and accurate quantitation of cis and trans FAME isomers.

Chromatographic Compression and Spectral Deconvolution

Unique VUV absorbance spectra not only enable unambiguous compound identification, and allows GC run times to be deliberately shortened. VUV detectors operate at ambient pressure and are thus not flow rate limited. GC run times can be reduced by increasing the GC column flow and oven temperature program rates.

Flow rate-enhanced chromatographic compression utilizes VUV spectral deconvolution to resolve any co-elution that may result from shortening GC runtimes. VUV absorption is additive, meaning that overlapping peaks give a spectrum that corresponds to the sum absorbance of each compound. The individual contribution of each analyte can be determined if the VUV spectra for co-eluting compounds are stored in the VUV library. The ability to differentiate coeluting analyte spectra and use them to deconvolve the overlapping signals is demonstrated in figure. The individual spectra of terpenes limonene and p-Cymene are shown in Panel A along with the summed absorbance of the selected retention time window (blue region in Panel B) and the fit with VUV library spectra. The R^2 >0.999 fit result confirms their identities, and enables the deconvolution of these and other terpenes analyzed by GC-VUV as featured in Panel B.

Testing for the presence of residual solvents in Active Pharmaceutical Ingredients (APIs) is critical for patient safety and commonly follows United States Pharmacopeia (USP) Method <467> guidelines, or more broadly, International Council for Harmonization (ICH) Guideline Q3C(R6).

The gas chromatography (GC) runtime suggested by USP Method 467 is approximately 60 min. A generic method for residual solvent analysis by GC-MS describes conditions that include a runtime of approximately 30 minutes. A GC-VUV and static headspace method was developed using a chromatographic compression strategy that resulted in a GC runtime of 8 minutes. The GC-VUV method uses a flow rate of 4 mL/min and an oven ramp of 35 °C (held for 1 min), followed by an increase to 245 °C at a rate of 30 °C/min.

Figure compares the results when the general conditions of the GC-MS method were followed against the GC-VUV method run with Class 2 residual solvents. Tetralin eluted at approximately 35 minutes using the GC-MS method conditions, whereas the analyte had a retention time of less than 7 minutes when the GC-VUV method was applied. The co-elution of m- and p-xylene occurred in both GC-MS and GC-VUV method runs. VUV software matched the analyte absorbance of both isomers with VUV library spectra to deconvolve the overlapping signals as displayed in figure. Goodness of fit information ensures that the correct compound assignment takes place during the post-run data analysis.

The flow rate-enhanced chromatographic compression strategy has been applied to a diverse set of applications since the development of the GC-VUV method for residual solvents analysis. The fast GC-VUV approach reduced GC runtimes for terpene analysis from 30 minutes to 9 minutes. It has also been demonstrated that GC runtimes as short as 14 minutes can be used for PIONA compound analysis of gasoline samples. Typical GC separation times range between 1 – 2 hours using alternative methods.

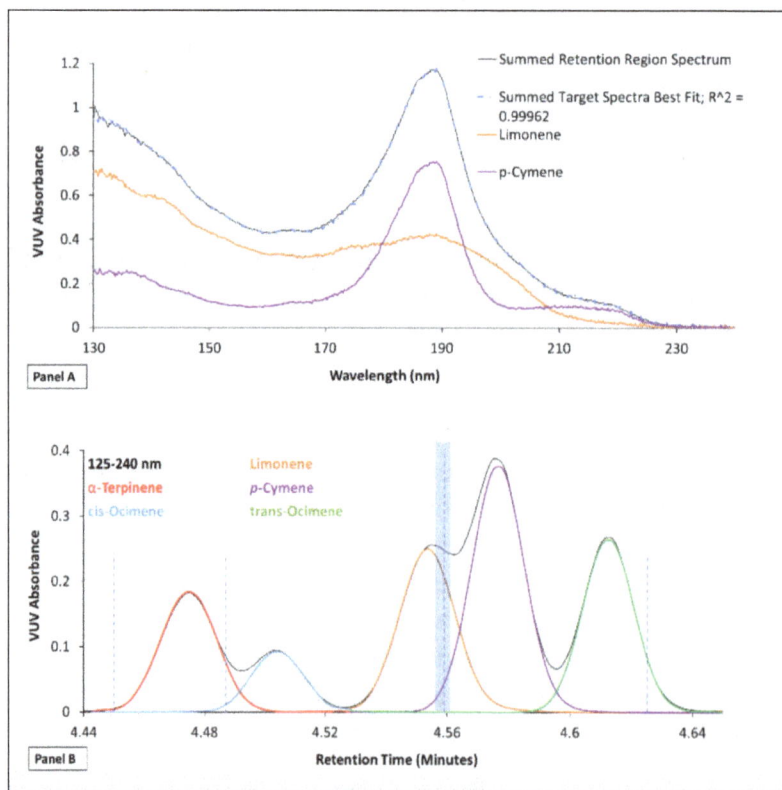

Panel A shows the individual spectra of limonene and p-Cymene along with the summed absorbance of the selected retention time (blue region in Panel B) and the fit with VUV library spectra. The deconvolution of these and other terpenes analyzed by fast GC-VUV is featured in Panel B.

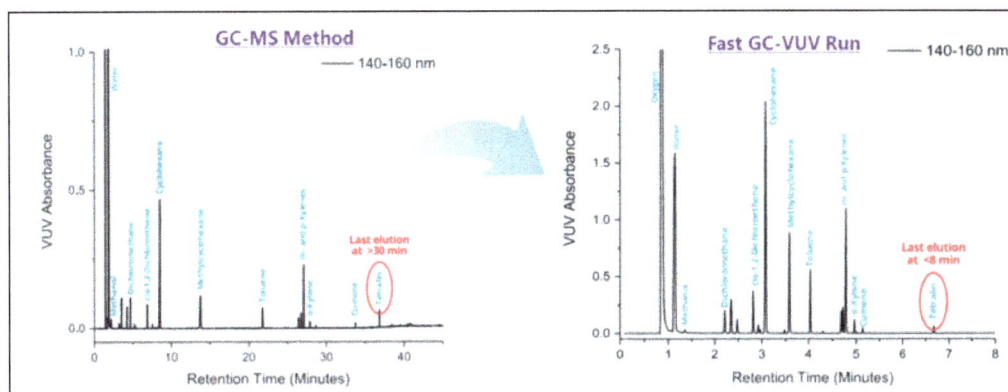

Comparison of legacy GC-MS and fast GC-VUV method runtimes for residual solvent analysis. Tetralin elutes at >30 minutes using the GC-MS method conditions, whereas the fast GC-VUV method elutes at <7 minutes.

Compound Class Characterization

GC-VUV can be used for bulk compositional analysis because compounds share spectral shape characteristics within a class. Proprietary software applies fitting procedures to quickly determine the relative contribution of each compound category present in a sample. Retention index information is used to limit the amount of VUV library searching and fitting performed for each analyte, enabling the automated data processing routine to be completed quickly. Compound class or specific compound concentrations can be reported as either mass or volume percent.

GC-VUV bulk compound characterization was first applied to the analysis of paraffin, isoparaffin, olefin, naphthene, and aromatic (PIONA) hydrocarbons in gasoline streams. It is suitable for use with finished gasoline, reformate, reformer feed, FCC, light naphtha, and heavy naphtha samples. A typical chromatographic analysis is displayed in figure. The inset shows how the analyte spectral response is fit with VUV library spectra for the selected time slice. A report detailing the carbon number breakdown within each PIONA compound class, as well as the relative mass or volume percent of classes, is shown. A table with mass % and carbon number data from a gasoline sample can be seen in figure. Compound class characterization utilizes a method known as time interval deconvolution (TID), which has recently been applied to the analysis of terpenes.

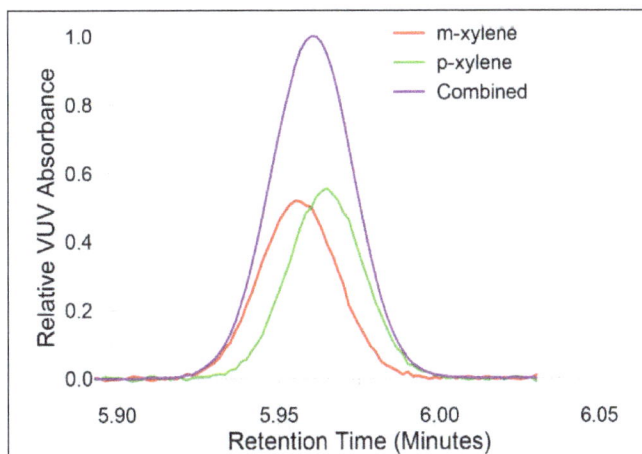

The deconvolution of m- and p-xylene co-elution. The relative contribution of each analyte is shown relative to the sum absorbance.

Headspace Gas Chromatography

Headspace gas chromatography uses headspace gas—from the top or "head" of a sealed container containing a liquid or solid brought to equilibrium—injected directly onto a gas chromatographic column for separation and analysis. In this process, only the most volatile (most readily existing as a vapor) substances make it to the column. The technique is commonly applied to the analysis of polymers, food and beverages, blood alcohol levels, environmental variables, cosmetics, and pharmaceutical ingredients.

Chemists often use the phrase "standard temperature and pressure" or "STP" to convey that they are working at a temperature of 25 °C and one atmosphere of pressure. There are three states of matter under these conditions: solids, liquids, and gases. Although all three are distinct states, both solids and gases can dissolve (or disperse) in liquids. The most commonly occurring liquid in the biosphere is water. All components of the atmosphere are capable of dissolving in water to some degree. The bulk of the stable natural components of the atmosphere are nitrogen, oxygen, carbon dioxide, gaseous water, argon, and other trace gases.

Materials that exist primarily in the gas phase at STP (i.e., "evaporates more than 95% by weight within six months under ambient evaporation testing conditions") are referred to as "volatile." Many natural and man-made (anthropogenic) materials are stable in two states at STP, earning them the title "semivolatile." A naturally occurring volatile that is sometimes found in aqueous solution is methane; water itself is semivolatile. Man-made or anthropogenic chemicals also occur in these classes. Examples of volatile anthropogenic chemicals include the refrigerants chlorofluorocarbons (CFCs) and hydrofluorocarbons (HCFCs). Semivolatile anthropogenics can exist as mixtures, such as petroleum distillates or as pure chemicals like trichloroethylene (TCE).

There is a need to analyze the dissolved gas content of aqueous solutions. Dissolved gases can directly interact with aquatic organisms or can volatilize from solution (the latter described by Henry's law). These processes can result in exposure that, depending on the nature of the dissolved material, can have negative health effects. There is natural occurrence of various dissolved gases in groundwater and can be a measure of health for lakes, streams, and rivers. Dissolved gases also occur as a result of human contamination from fuel and chlorinated spill sites. As such, headspace gas chromatography offers a method for determining if there is natural biodegradation processes occurring in contaminated aquifers. For example, fuel hydrocarbons will break down into methane. Chlorinated solvents such as trichloroethylene, break down into ethene and chloride. Detecting these compounds can determine if biodegradation processes are occurring and possibly at what rate. Natural gas extracted from the earth also contains many low molecular weight hydrocarbon compounds such as methane, ethane, propane, and butane. For example, methane has been found in many water wells in West Virginia.

RSKSOP-175 Analysis Method

One of the most widely used methods for headspace analysis is described by the United States Environmental Protection Agency (USEPA). Originally developed by the R.S. Kerr USEPA Laboratory in Ada, Oklahoma as a "high quality, defendable, and documented way to measure" methane, ethane, and ethene, RSKSOP-175 is a standard operating procedure (SOP) and an unofficial method employed by the USEPA to detect and quantify dissolved gases in water. This method has been used to quantify dissolved hydrogen, methane, ethylene, ethane, propane, butane, acetylene, nitrogen, nitrous oxide, and oxygen. The method uses headspace gas injected into a gas chromatographic column (GC) to determine the original concentration in a water sample.

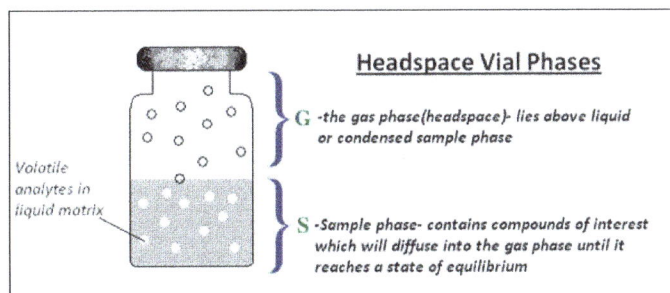

Phases of a headspace vial used in gas chromatography.

A sample of water is collected in the field in a vial without headspace and capped with a Teflon septum or crimp top to minimize the escape of volatile gases. It is beneficial to store the bottles upside down to further minimize loss of analytes. Before analysis begins, the sample is brought to room temperature and temperature is recorded. In the laboratory, a headspace is created by displacing water with high purity helium. The bottle is then shaken upside down for a minimum of five minutes in order to equilibrate the dissolved gases into the headspace. It's important to note that the bottle must be kept upside down for the remainder of analysis if manually injected. A known volume of headspace gas is then injected onto a gas chromatographic column. An automated process can also be utilized. Individual components (gases) are separated and detected by either a thermal conductivity detector (TCD), a flame ionization detector (FID), or an electron capture detector (ECD). Using the known temperature of the sample, the bottle volume, the concentrations of gas in the headspace (as determined by GC), and Henry's law constant, the concentration of the original water sample is calculated.

Calculations

Using the known temperature of the sample, the bottle volume, the concentrations of gas in the headspace (as determined by GC), and Henry's law constant, the concentration of the original water sample is calculated. Total gas concentration (TC) in the original water sample is calculated by determining the concentration of headspace and converting this to the partial pressure and then solving for the aqueous concentration which partitioned in the gas phase (C_{AH}) and the concentration remaining in the aqueous phase (C_A). The total concentration of gas in original sample (TC) is the sum of the concentration partitioned in the gas phase (C_{AH}) and the concentration remaining in the aqueous phase (C_A):

$$TC = C_{AH} + C_A$$

Henry's law states that the mole fraction of a dissolved gas (x_g) is equal to the partial pressure of the gas (p_g) at equilibrium divided by Henry's law constant (H). Gas solubility coefficients are used to calculate Henry's law constant:

$$x_g = p_g / H$$

After manipulating equations and substituting volumes of each phase, the molar concentration of water (55.5 mol/L) and the molecular weight of the gas analyte (MW), a final equation is solved:

$$TC = (55.5mol / L) * p_g / H * MW(g / mol) * 10^3 mg / g + [(V_h / (V_b - V_h)] * C_g *$$

$$(MW(g / mol) / (22.4L / mol)) * [273K / (T + 273K)] * 10^3 mg / g$$

Where V_b is the bottle volume and V_h is the volume of headspace. C_g is the volumetric concentration of gas. For full calculation examples, reference RSK-175SOP.

Practical Considerations

One of the major concerns for this method is reproducibility. Due to the nature of calculations, this method is reliant on temperatures to be constant and volumes to be exact. When gases are spiked manually into the GC, the speed and technique in which an analyst does this plays a role in reproducibility. If one analyst is faster in removing gas from the vial and injecting it onto the instrument, then it is important to have the same analyst run on the calibration they prepped, otherwise error will more than likely be introduced. A headspace auto-sampler may remove some of this error, but constant heat and variable temperature on the instrument becomes an issue.

Other Methods and Techniques

Prior to RSKSOP-175, the EPA used Method 3810, which before that was Method 5020. However, Method 3810 is still used by some laboratories.

Other headspace GC methods include:

- ASTM D4526-12 and ASTM D8028-17.

- EPA 5021A.

- Pennsylvania Department of Environmental Protection (PA-DEP) 3686 (#BOL 6019).

Preparative Gas Chromatography

A simple, convenient, and highly efficient preparative GC system has been developed that uses short sections of megabore capillary columns as sample collection (sorbent) traps. The performance of this system with various types of capillary column traps and under various collection conditions was systematically investigated with model compounds, including C4 to C20 normal alkanes, esters, and alcohols. The thickness and polarity of the sorptive stationary phase and the temperature of the collection trap affected trap performance. Each group of compounds was efficiently trapped above a critical Kovat's index, and the type of trap (deactivated, methyl polysiloxane, polyethylene glycol), film thickness, and whether or not the trap was cooled significantly shifted this threshold index. Above this critical index, recovery efficiencies of traps with methyl polysiloxane films were 80-100% for a wide range of injected sample mass. For example, a DB-1 collection trap with a film thickness of 1.5 microm methyl polysiloxane operated at ambient temperature trapped >84% of the mass of injected compounds of all three chemical classes with Kovat's index >1,100 (determined on a nonpolar column) with injected sample mass ranging from 10 to 1,000 ng of each compound. This preparative GC system is technically and economically feasible for most researchers. Furthermore, it is suitable for the preparation of NMR samples of volatile and semivolatile compounds, especially with sample sizes ranging from several nanograms to several micrograms.

In Pharmaceutical Analysis

The major success of the application of GC in pharmaceutical quantitative analysis is firstly due to the very high efficiencies of separation power, secondly to the extreme sensitivity of the detection of even very small amounts of separated species and finally to the precision and accuracy of the data from quantitative analyses of very complex mixtures. GC analyses are also easy to automate from sample introduction to separation. Because of the main advantages and its short analysis time and reliable results GC is used as quality control purposes in the pharmaceutical industry. In fact pharmaceutical analysis generally involves two steps; separation of the compound of interest and quantitation of the compounds. The better the separation the easier the quantitation. GC detectors have different responses to each compound. In order to determine quantitative amounts of various compounds in a separation the detector must be calibrated using standards. Standard solutions of sample are injected and the detector response recorded. Comparison of the standard and sample retention times allows qualitative analysis of the sample. Comparison of the peak area of the standards with that of the sample allows quantitation of analyte. Due to this fact, GC is widely used as a routine analytical technique in pharmaceutical quantitative analysis mostly used in for the determination of organic volatile impurities and nicotine level during drugs formulation.

Determination of Organic Volatile Impurities by GC

Organic Volatile impurities are residual solvents that are used in and are produced during the synthesis of drug substances, or in excipients used in the production of drug formulations. Many of these residual solvents generally cannot be completely removed by standard manufacturing processes or techniques and are left behind, preferably at low levels. Organic solvents such as acetone, ethyl acetate, isopropyl alcohol, methanol, tetrahydrofuran and toluene frequently used in pharmaceutical industry for the manufacturing of Active Pharmaceutical ingredients therefore, in manufacturing drug substances and from one or more steps of the synthetic process, volatile solvents can be retained in the end products. Most of the time ethanol and acetone are used in the preparation of the polymeric coating of tablets. On other hand isopropyl alcohol is used in the crystallization of the active ingredient while ethyl acetate is a process solvent for the gel forming polymer. Low levels of these organic solvents are inevitably present in the drug product even after drying process. These organic volatile residuals affect physiochemical properties of a drug, such as particle size, dissolution rate and stability, but also can present a serious potential health hazard. Very often these solvents, referred to as residual solvents, are carried to the pharmaceutical preparation concerned and making their determination very important. Therefore, GC is superior to other techniques for analysis of these residual solvents. It provide good retention and separation at low oven temperatures. Due to the fact the content of residual organic solvents in pharmaceutical industry is routinely measured by GC technique.

Determination of Nicotine by GC

Because of its rapid and accurate analytical result; GC is used to determine the nicotine level in pharmaceutical drugs formulation. GC applications in combination with other techniques are also vital in pharmaceutical industries for isolation and characterization of volatile compounds. Currently the use of GC in pharmaceutical quantitative analysis is very usual and include the analysis of samples of active pharmaceutical ingredients and their intermediates as well as in-process testing for residual solvents to optimize the drying process.

Disadvantages and Advantages of GC

Gas chromatography (GC) is an analytical method which is used for the separation of an volatile substance from a give mixture of compounds which are very difficult to separate and analyse.

This type of chromatography separates the molecules based on the volatility of a substance. The substance with more volatility will separate out quickly while the substance with less volatility will elute out slowly.

Advantages

- This method has a high resolution power compared to other methods.

- This method has high sensitivity when used with thermal detectors.

- This technique gives relatively good accuracy and precision.

- Separation and analysis of sample very quickly.

- Sample with less quantity is also separated.

Disadvantages

- Only volatile samples or the sample which can be made volatile are separated by this method.

- During injection of the gaseous sample proper attention is required.

- The sample of gas which is about to inject must be thermally stable so that it does not get degraded when heated.

References

- Gas-Chromatography, Chromatography, Instrumental-Analysis, Supplemental-Modules-(Analytical-Chemistry), Bookshelves: libretexts.org, Retrieved 1 January, 2019

- Carvalho, Matheus (2018). "Osmar, the open-source microsyringe autosampler". Hardwarex. 3: 10–38. Doi:10.1016/j.ohx.2018.01.001

- Detectors, moreinfo: 50megs.com, Retrieved 13 March, 2019

- Sithersingh, M.J.; Snow, N.H. (2012). "Chapter 9: Headspace-Gas Chromatography". In Poole, C. (ed.). Gas Chromatography. Elsevier. Pp. 221–34. ISBN 9780123855404

- Gaschrm, chrom, tutorials, chemistry, hwb: shu.ac.uk, Retrieved 3 February, 2019

- USGS. "Lower Columbia River Dissolved Gas Monitoring Network". Oregon Water Science Center. Retrieved 16 April 2019

- Gas-chromatography: microbenotes.com, Retrieved 26 August, 2019

- Quantitative-use-of-gas-chromotography-in-pharmaceutical-analysis-biology-essay, biology, essays: ukes-says.com, Retrieved 5 April, 2019

- Disadvantages-advantages-of-gc: frndzzz.com, Retrieved 13 February, 2019

Liquid Chromatography 3

The analytical chromatographic technique that is used to separate ions and molecules that are dissolved in a liquid mobile phase is referred to as liquid chromatography. The usage of high pressure for separation, detection and quantification of different components in a mixture is termed as high performance liquid chromatography. The topics elaborated in this chapter will help in gaining a better perspective about the types and applications of high performance liquid chromatography.

Liquid chromatography (LC) is an analytical chromatographic technique that is useful for separating ions or molecules that are dissolved in a solvent.

If the sample solution is in contact with a second solid or liquid phase, the different solutes will interact with the other phase to differing degrees due to differences in adsorption, ion-exchange, partitioning, or size.

These differences allow the mixture components to be separated from each other by using these differences to determine the transit time of the solutes through a column.

Instrumentation Simple liquid chromatography consists of a column with a fritted bottom that holds a stationary phase in equilibrium with a solvent. Typical stationary phases (and their interactions with the solutes) are: Solids (adsorption), ionic groups on a resin (ion-exchange), liquids on an inert solid support (partitioning), and porous inert particles (size-exclusion).

The mixture to be separated is loaded onto the top of the column followed by more solvent. The different components in the sample mixture pass through the column at different rates due to differences in their partitioning behavior between the mobile liquid phase and the stationary phase. The compounds are separated by collecting aliquots of the column effluent as a function of time.

Conventional LC is most commonly used in preparative scale work to purify and isolate some components of a mixture. It is also used in ultratrace separations where small disposable columns are used once and then discarded. Analytical separations of solutions for detection or quantification typically use more sophisticated high-performance liquid chromatography instruments.

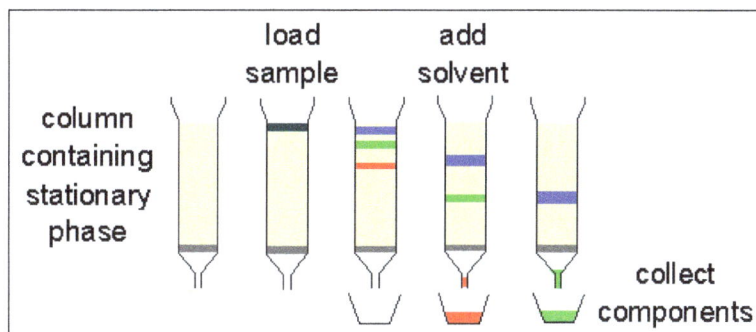

Schematic of a simple liquid chromatographic separation.

HPLC instruments use a pump to force the mobile phase through and provide higher resolution and faster analysis time.

High Performance Liquid Chromatography (HPLC)

High-performance liquid chromatography (HPLC; formerly referred to as high-pressure liquid chromatography) is a technique in analytical chemistry used to separate, identify, and quantify each component in a mixture. It relies on pumps to pass a pressurized liquid solvent containing the sample mixture through a column filled with a solid adsorbent material. Each component in the sample interacts slightly differently with the adsorbent material, causing different flow rates for the different components and leading to the separation of the components as they flow out of the column.

HPLC has been used for manufacturing (e.g., during the production process of pharmaceutical and biological products), legal (e.g., detecting performance enhancement drugs in urine), research (e.g., separating the components of a complex biological sample, or of similar synthetic chemicals from each other), and medical (e.g., detecting vitamin D levels in blood serum) purposes.

Chromatography can be described as a mass transfer process involving adsorption. HPLC relies on pumps to pass a pressurized liquid and a sample mixture through a column filled with adsorbent, leading to the separation of the sample components. The active component of the column, the adsorbent, is typically a granular material made of solid particles (e.g., silica, polymers, etc.), 2–50 μm in size. The components of the sample mixture are separated from each other due to their different degrees of interaction with the adsorbent particles. The pressurized liquid is typically a mixture of solvents (e.g., water, acetonitrile and/or methanol) and is referred to as a "mobile phase". Its composition and temperature play a major role in the separation process by influencing the interactions taking place between sample components and adsorbent. These interactions are physical in nature, such as hydrophobic (dispersive), dipole–dipole and ionic, most often a combination.

HPLC is distinguished from traditional ("low pressure") liquid chromatography because operational pressures are significantly higher (50–350 bar), while ordinary liquid chromatography typically relies on the force of gravity to pass the mobile phase through the column. Due to the small sample amount separated in analytical HPLC, typical column dimensions are 2.1–4.6 mm diameter, and 30–250 mm length. Also HPLC columns are made with smaller adsorbent particles (2–50 μm in average particle size). This gives HPLC superior resolving power (the ability to distinguish between compounds) when separating mixtures, which makes it a popular chromatographic technique.

The schematic of a HPLC instrument typically includes a degasser, sampler, pumps, and a detector. The sampler brings the sample mixture into the mobile phase stream which carries it into the column. The pumps deliver the desired flow and composition of the mobile phase through the column. The detector generates a signal proportional to the amount of sample component emerging from the column, hence allowing for quantitative analysis of the sample components. A digital microprocessor and user software control the HPLC instrument and provide data analysis. Some models of mechanical pumps in a HPLC instrument can mix multiple solvents together in ratios changing in time, generating a composition gradient in the mobile phase. Various detectors are in common use, such as UV/Vis, photodiode array (PDA) or based on mass spectrometry. Most HPLC instruments also have a column oven that allows for adjusting the temperature at which the separation is performed.

A modern self-contained HPLC.

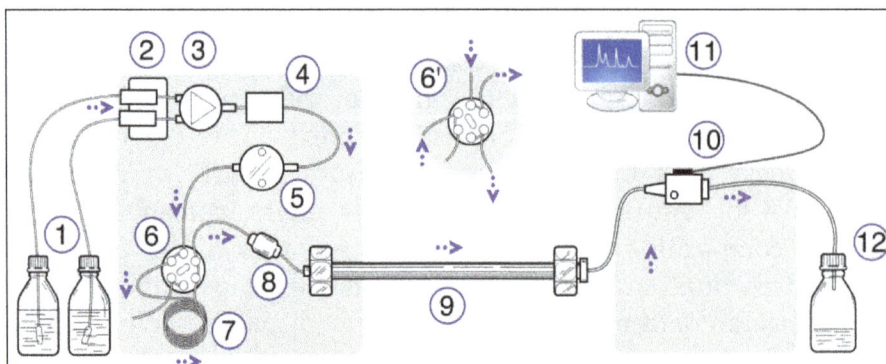

Schematic representation of an HPLC unit: (1) Solvent reservoirs, (2) Solvent degasser, (3) Gradient valve, (4) Mixing vessel for delivery of the mobile phase, (5) High-pressure pump, (6) Switching valve in "inject position", (6') Switching valve in "load position", (7) Sample injection loop, (8) Pre-column (guard column), (9) Analytical column, (10) Detector (i.e., IR, UV), (11) Data acquisition, (12) Waste or fraction collector.

Operation

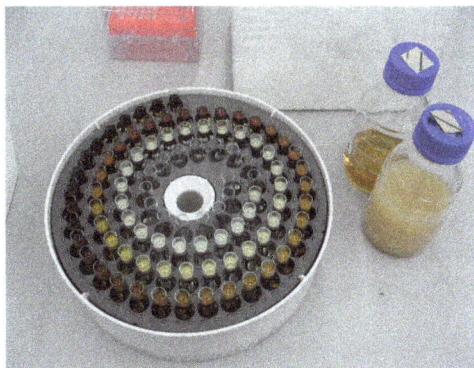

A rotary fraction collector collecting HPLC output. The system is being used to isolate a fraction containing Complex I from E. coli plasma membranes. About 50 litres of bacteria were needed to isolate this amount.

The sample mixture to be separated and analyzed is introduced, in a discrete small volume (typically microliters), into the stream of mobile phase percolating through the column. The components of the sample move through the column at different velocities, which are a function of specific physical interactions with the adsorbent (also called stationary phase). The velocity of each component depends on its chemical nature, on the nature of the stationary phase (column) and on the composition of the mobile phase. The time at which a specific analyte elutes (emerges from the column) is called its retention time. The retention time measured under particular conditions is an identifying characteristic of a given analyte.

Many different types of columns are available, filled with adsorbents varying in particle size, and in the nature of their surface ("surface chemistry"). The use of smaller particle size packing materials requires the use of higher operational pressure ("backpressure") and typically improves chromatographic resolution (the degree of peak separation between consecutive analytes emerging from the column). Sorbent particles may be hydrophobic or polar in nature.

Common mobile phases used include any miscible combination of water with various organic solvents (the most common are acetonitrile and methanol). Some HPLC techniques use water-free mobile phases. The aqueous component of the mobile phase may contain acids (such as formic, phosphoric or trifluoroacetic acid) or salts to assist in the separation of the sample components. The composition of the mobile phase may be kept constant ("isocratic elution mode") or varied ("gradient elution mode") during the chromatographic analysis. Isocratic elution is typically effective in the separation of sample components that are very different in their affinity for the stationary phase. In gradient elution the composition of the mobile phase is varied typically from low to high eluting strength. The eluting strength of the mobile phase is reflected by analyte retention times with high eluting strength producing fast elution (=short retention times). A typical gradient profile in reversed phase chromatography might start at 5% acetonitrile (in water or aqueous buffer) and progress linearly to 95% acetonitrile over 5–25 minutes. Periods of constant mobile phase composition may be part of any gradient profile. For example, the mobile phase composition may be kept constant at 5% acetonitrile for 1–3 min, followed by a linear change up to 95% acetonitrile.

The chosen composition of the mobile phase (also called eluent) depends on the intensity of interactions between various sample components ("analytes") and stationary phase (e.g., hydrophobic interactions in reversed-phase HPLC). Depending on their affinity for the stationary and mobile phases analytes partition between the two during the separation process taking place in the column. This partitioning process is similar to that which occurs during a liquid–liquid extraction but is continuous, not step-wise. In this example, using a water/acetonitrile gradient, more hydrophobic components will elute (come off the column) late, once the mobile phase gets more concentrated in acetonitrile (i.e., in a mobile phase of higher eluting strength).

The choice of mobile phase components, additives (such as salts or acids) and gradient conditions depends on the nature of the column and sample components. Often a series of trial runs is performed with the sample in order to find the HPLC method which gives adequate separation.

High Perfomance Liquid Chromatography

High performance liquid chromatography (HPLC) is basically a highly improved form of column liquid chromatography.

Instead of a solvent being allowed to drip through a column under gravity, it is forced through under high pressures of up to 400 atmospheres. That makes it much faster.

All chromatographic separations, including HPLC operate under the same basic principle; separation of a sample into its constituent parts because of the difference in the relative affinities of different molecules for the mobile phase and the stationary phase used in the separation.

Types of HPLC

There are following variants of HPLC, depending upon the phase system (stationary) in the process:

Normal Phase HPLC

This method separates analytes on the basis of polarity. NP-HPLC uses polar stationary phase and non-polar mobile phase. Therefore, the stationary phase is usually silica and typical mobile phases are hexane, methylene chloride, chloroform, diethyl ether, and mixtures of these.

Polar samples are thus retained on the polar surface of the column packing longer than less polar materials.

Reverse Phase HPLC

The stationary phase is nonpolar (hydrophobic) in nature, while the mobile phase is a polar liquid, such as mixtures of water and methanol or acetonitrile. It works on the principle of hydrophobic interactions hence the more nonpolar the material is, the longer it will be retained.

Size-exclusion HPLC

The column is filled with material having precisely controlled pore sizes, and the particles are separated according to its their molecular size. Larger molecules are rapidly washed through the column; smaller molecules penetrate inside the porous of the packing particles and elute later.

Ion-Exchange HPLC

The stationary phase has an ionically charged surface of opposite charge to the sample ions. This technique is used almost exclusively with ionic or ionizable samples.

The stronger the charge on the sample, the stronger it will be attracted to the ionic surface and thus, the longer it will take to elute. The mobile phase is an aqueous buffer, where both pH and ionic strength are used to control elution time.

Instrumentation of HPLC

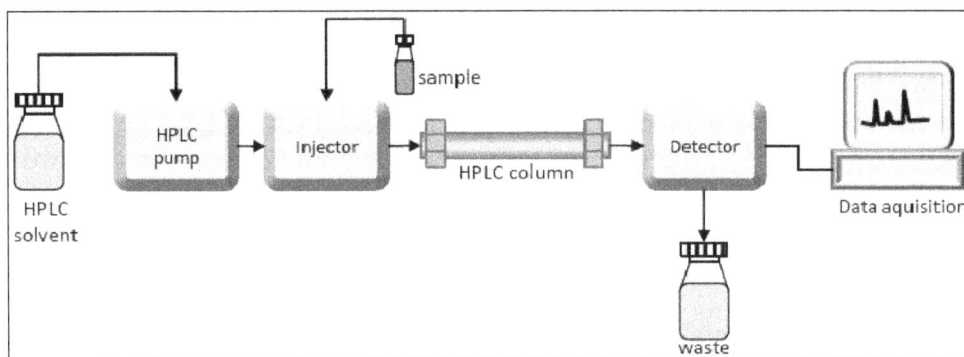

HPLC instrumentation includes a pump, injector, column, detector and integrator or acquisition and display system. The heart of the system is the column where separation occurs.

Solvent Reservoir

Mobile phase contents are contained in a glass reservoir. The mobile phase, or solvent, in HPLC is usually a mixture of polar and non-polar liquid components whose respective concentrations are varied depending on the composition of the sample.

Pump

A pump aspirates the mobile phase from the solvent resorvoir and forces it through the system's column and detecter. Depending on a number of factors including column dimensions, particle size of the stationary phase, the flow rate and composition of the mobile phase, operating pressures of up to 42000 kPa (about 6000 psi) can be generated.

Sample Injector

The injector can be a single injection or an automated injection system. An injector for an HPLC system should provide injection of the liquid sample within the range of 0.1-100 mL of volume with high reproducibility and under high pressure (up to 4000 psi).

Columns

Columns are usually made of polished stainless steel, are between 50 and 300 mm long and have an internal diameter of between 2 and 5 mm. They are commonly filled with a stationary phase with a particle size of 3–10 μm.

Columns with internal diameters of less than 2 mm are often referred to as microbore columns. Ideally the temperature of the mobile phase and the column should be kept constant during an analysis.

Detector

The HPLC detector, located at the end of the column detect the analytes as they elute from the chromatographic column. Commonly used detectors are UV-spectroscopy, fluorescence, mass-spectrometric and electrochemical detectors.

Data Collection Devices

Signals from the detector may be collected on chart recorders or electronic integrators that vary in complexity and in their ability to process, store and reprocess chromatographic data. The computer integrates the response of the detector to each component and places it into a chromatograph that is easy to read and interpret.

Applications of HPLC

The information that can be obtained by HPLC includes resolution, identification and quantification of a compound. It also aids in chemical separation and purification. The other applications of HPLC include:

- Pharmaceutical Applications:
  - To control drug stability.
  - Tablet dissolution study of pharmaceutical dosages form.
  - Pharmaceutical quality control.
- Environmental Applications:
  - Detection of phenolic compounds in drinking water.
  - Bio-monitoring of pollutants.
- Applications in Forensics:
  - Quantification of drugs in biological samples.
  - Identification of steroids in blood, urine etc.
  - Forensic analysis of textile dyes.
  - Determination of cocaine and other drugs of abuse in blood, urine etc.
- Food and Flavour:
  - Measurement of Quality of soft drinks and water.
  - Sugar analysis in fruit juices.

◦ Analysis of polycyclic compounds in vegetables.

◦ Preservative analysis.

- Applications in Clinical Tests:

    ◦ Urine analysis, antibiotics analysis in blood.

    ◦ Analysis of bilirubin, biliverdin in hepatic disorders.

    ◦ Detection of endogenous Neuropeptides in extracellular fluid of brain etc.

Principle of HPLC

The separation principle of HPLC is based on the distribution of the analyte (sample) between a mobile phase (eluent) and a stationary phase (packing material of the column). Depending on the chemical structure of the analyte, the molecules are retarded while passing the stationary phase. The specific intermolecular interactions between the molecules of a sample and the packing material define their time "on-column". Hence, different constituents of a sample are eluted at different times. Thereby, the separation of the sample ingredients is achieved.

A detection unit (e.g. UV detector) recognizes the analytes after leaving the column. The signals are converted and recorded by a data management system (computer software) and then shown in a chromatogram. After passing the detector unit, the mobile phase can be subjected to additional detector units, a fraction collection unit or to the waste. In general, a HPLC system contains the following modules: a solvent reservoir, a pump, an injection valve, a column, a detector unit and a data processing unit. The solvent (eluent) is delivered by the pump at high pressure and constant speed through the system. To keep the drift and noise of the detector signal as low as possible, a constant and pulseless flow from the pump is crucial. The analyte (sample) is provided to the eluent by the injection valve.

Working Principle of High Performance Liquid Chromatography

A reservoir holds the solvent called the mobile phase, because it moves. A high-pressure pump solvent delivery system or solvent manager is used to generate and meter a specified flow rate of mobile phase, typically milliliters per minute. An injector sample manager or autosampler is able to introduce inject the sample into the continuously flowing mobile phase stream that carries the sample into the HPLC column. The column contains the chromatographic packing material needed to effect the separation. This packing material is called the stationary phase because it is held in place by the column hardware.

A detector is needed to see the separated compound bands as they elute from the HPLC column most compounds have no color, so we cannot see them with our eyes. The mobile phase exits the detector and can be sent to waste, or collected, as desired. When the mobile phase contains a separated compound band, HPLC provides the ability to collect this fraction of the eluate containing that purified compound for further study. This is called preparative chromatography.

High-pressure tubing and fittings are used to interconnect the pump, injector, column, and detector components to form the conduit for the mobile phase, sample, and separated compound bands.

The detector is wired to the computer data station, the HPLC system component that records the electrical signal needed to generate the chromatogram on its display and to identify and quantitate the concentration of the sample constituents. Since sample compound characteristics can be very different, several types of detectors have been developed. For example, if a compound can absorb ultraviolet light, a UV-absorbance detector is used. If the compound fluoresces, a fluorescence detector is used. If the compound does not have either of these characteristics, a more universal type of detector is used, such as an evaporative-light-scattering detector [ELSD]. The most powerful approach is the use multiple detectors in series. For example, a UV and/or ELSD detector may be used in combination with a mass spectrometer [MS] to analyze the results of the chromatographic separation. This provides, from a single injection, more comprehensive information about an analyte. The practice of coupling a mass spectrometer to an HPLC system is called LC/MS.

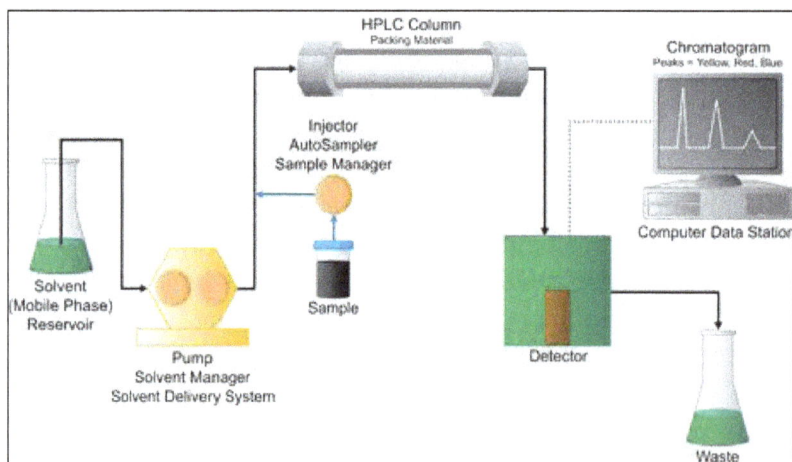

High-Performance Liquid Chromatography [HPLC] System.

HPLC Operation

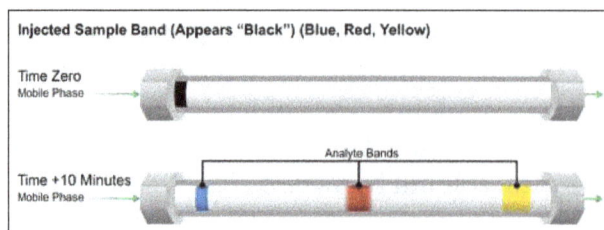

Mobile phase enters the column from the left, passes through the particle bed, and exits at the right. Flow direction is represented by green arrows. The sample shown here, a mixture of yellow, red, and blue dyes, appears at the inlet of the column as a single black band. In reality, this sample could be anything that can be dissolved in a solvent; typically the compounds would be colorless and the column wall opaque, so we would need a detector to see the separated compounds as they elute.

After a few minutes, during which mobile phase flows continuously and steadily past the packing material particles, we can see that the individual dyes have moved in separate bands at different speeds. This is because there is a competition between the mobile phase and the stationary phase for attracting each of the dyes or analytes. Notice that the yellow dye band moves the fastest and is about to exit the column. The yellow dye likes [is attracted to] the mobile phase more than the other dyes. Therefore, it moves at a faster speed, closer to that of the mobile phase. The blue dye band likes the packing material more than the mobile phase. Its stronger attraction to the particles causes it to move significantly slower. In other words, it is the most retained compound in this sample mixture. The red dye band has an intermediate attraction for the mobile phase and therefore moves at an intermediate speed through the column. Since each dye band moves at different speed, we are able to separate it chromatographically.

Detector

As the separated dye bands leave the column, they pass immediately into the detector. The detector contains a flow cell that sees [detects] each separated compound band against a background of mobile phase. In reality, solutions of many compounds at typical HPLC analytical concentrations are colorless. An appropriate detector has the ability to sense the presence of a compound and send its corresponding electrical signal to a computer data station. A choice is made among many different types of detectors, depending upon the characteristics and concentrations of the compounds that need to be separated and analysed.

Types of HPLC

Prior to HPLC scientists used standard liquid chromatographic techniques. Liquid chromatographic systems were largely inefficient due to the flow rate of solvents being dependent on gravity. Separations took many hours, and sometimes days to complete. Gas chromatography (GC) at the time was more powerful than liquid chromatography (LC), however, it was believed that gas phase separation and analysis of very polar high molecular weight biopolymers was impossible. GC was ineffective for many biochemists because of the thermal instability of the solutes. As a result, alternative methods were hypothesized which would soon result in the development of HPLC.

Following on the seminal work of Martin and Synge in 1941, it was predicted by Cal Giddings, Josef Huber, and others in the 1960s that LC could be operated in the high-efficiency mode by reducing the packing-particle diameter substantially below the typical LC (and GC) level of 150 μm and using pressure to increase the mobile phase velocity. These predictions underwent extensive experimentation and refinement throughout the 60s into the 70s. Early developmental research began to improve LC particles, and the invention of Zipax, a superficially porous particle, was promising for HPLC technology.

The 1970s brought about many developments in hardware and instrumentation. Researchers began using pumps and injectors to make a rudimentary design of an HPLC system. Gas amplifier pumps were ideal because they operated at constant pressure and did not require leak free seals or check valves for steady flow and good quantitation. Hardware milestones were made at Dupont IPD (Industrial Polymers Division) such as a low-dwell-volume gradient device being utilized as well as replacing the septum injector with a loop injection valve.

While instrumentational developments were important, the history of HPLC is primarily about the history and evolution of particle technology. After the introduction of porous layer particles, there has been a steady trend to reduced particle size to improve efficiency. However, by decreasing particle size, new problems arose. The practical disadvantages stem from the excessive pressure drop needed to force mobile fluid through the column and the difficulty of preparing a uniform packing of extremely fine materials. Every time particle size is reduced significantly, another round of instrument development usually must occur to handle the pressure.

Partition Chromatography

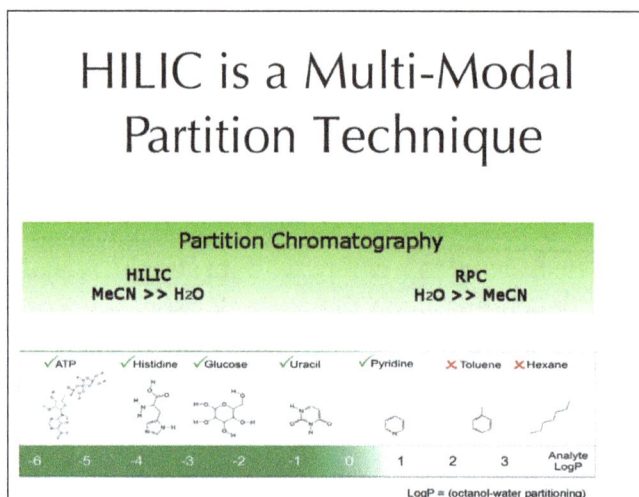

HILIC Partition Technique Useful Range.

Partition chromatography was one of the first kinds of chromatography that chemists developed. The partition coefficient principle has been applied in paper chromatography, thin layer chromatography, gas phase and liquid–liquid separation applications. The 1952 Nobel Prize in chemistry was earned by Archer John Porter Martin and Richard Laurence Millington Synge for their development of the technique, which was used for their separation of amino acids. Partition chromatography uses a retained solvent, on the surface or within the grains or fibers of an "inert" solid supporting matrix as with paper chromatography; or takes advantage of some coulombic and/or hydrogen donor interaction with the stationary phase. Analyte molecules partition between a liquid stationary phase and the eluent. Just as in Hydrophilic Interaction Chromatography (HILIC; a sub-technique within HPLC), this method separates analytes based on differences in their polarity. HILIC most often uses a bonded polar stationary phase and a mobile phase made primarily of acetonitrile with water as the strong component. Partition HPLC has been used historically on unbonded silica or alumina supports. Each works effectively for separating analytes by relative polar differences. HILIC bonded phases have the advantage of separating acidic, basic and neutral solutes in a single chromatographic run.

The polar analytes diffuse into a stationary water layer associated with the polar stationary phase and are thus retained. The stronger the interactions between the polar analyte and the polar stationary phase (relative to the mobile phase) the longer the elution time. The interaction strength depends on the functional groups part of the analyte molecular structure, with more polarized groups (e.g., hydroxyl-) and groups capable of hydrogen bonding inducing more retention. Coulombic (electrostatic) interactions can also increase retention. Use of more polar solvents in the mobile phase will decrease the retention time of the analytes, whereas more hydrophobic solvents tend to increase retention times.

Normal–phase Chromatography

Normal–phase chromatography was one of the first kinds of HPLC that chemists developed. Also known as normal-phase HPLC (NP-HPLC) this method separates analytes based on their affinity for a polar stationary surface such as silica, hence it is based on analyte ability to engage in polar interactions (such as hydrogen-bonding or dipole-dipole type of interactions) with the sorbent surface. NP-HPLC uses a non-polar, non-aqueous mobile phase (e.g., Chloroform), and works effectively for separating analytes readily soluble in non-polar solvents. The analyte associates with and is retained by the polar stationary phase. Adsorption strengths increase with increased analyte polarity. The interaction strength depends not only on the functional groups present in the structure of the analyte molecule, but also on steric factors. The effect of steric hindrance on interaction strength allows this method to resolve (separate) structural isomers.

The use of more polar solvents in the mobile phase will decrease the retention time of analytes, whereas more hydrophobic solvents tend to induce slower elution (increased retention times). Very polar solvents such as traces of water in the mobile phase tend to adsorb to the solid surface of the stationary phase forming a stationary bound (water) layer which is considered to play an active role in retention. This behavior is somewhat peculiar to normal phase chromatography because it is governed almost exclusively by an adsorptive mechanism (i.e., analytes interact with a solid surface rather than with the solvated layer of a ligand attached to the sorbent surface). Adsorption chromatography is still widely used for structural isomer separations in both column and thin-layer chromatography formats on activated (dried) silica or alumina supports.

Partition- and NP-HPLC fell out of favor in the 1970s with the development of reversed-phase HPLC because of poor reproducibility of retention times due to the presence of a water or protic organic solvent layer on the surface of the silica or alumina chromatographic media. This layer changes with any changes in the composition of the mobile phase (e.g., moisture level) causing drifting retention times.

Recently, partition chromatography has become popular again with the development of Hilic bonded phases which demonstrate improved reproducibility, and due to a better understanding of the range of usefulness of the technique.

Displacement Chromatography

The basic principle of displacement chromatography is: A molecule with a high affinity for the chromatography matrix (the displacer) will compete effectively for binding sites, and thus displace all molecules with lesser affinities. There are distinct differences between displacement and elution chromatography. In elution mode, substances typically emerge from a column in narrow, Gaussian peaks. Wide separation of peaks, preferably to baseline, is desired in order to achieve maximum purification. The speed at which any component of a mixture travels down the column in elution mode depends on many factors. But for two substances to travel at different speeds, and thereby be resolved, there must be substantial differences in some interaction between the biomolecules and the chromatography matrix. Operating parameters are adjusted to maximize the effect of this difference. In many cases, baseline separation of the peaks can be achieved only with gradient elution and low column loadings. Thus, two drawbacks to elution mode chromatography, especially at the preparative scale, are operational complexity, due to gradient solvent pumping, and low throughput, due to low column loadings. Displacement chromatography has advantages over elution chromatography in that components are resolved into consecutive zones of pure substances rather than "peaks". Because the process takes advantage of the nonlinearity of the isotherms, a larger column feed can be separated on a given column with the purified components recovered at significantly higher concentration.

Reversed-phase Chromatography (RPC)

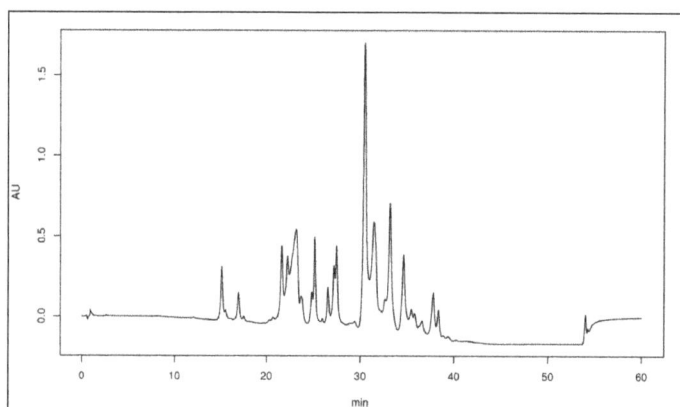

A chromatogram of complex mixture (perfume water) obtained by reversed phase HPLC.

Reversed phase HPLC (RP-HPLC) has a non-polar stationary phase and an aqueous, moderately polar mobile phase. One common stationary phase is a silica which has been surface-modified with RMe_2SiCl, where R is a straight chain alkyl group such as $C_{18}H_{37}$ or C_8H_{17}. With such stationary phases, retention time is longer for molecules which are less polar, while polar molecules elute more readily (early in the analysis). An investigator can increase retention times by adding more water to the mobile phase; thereby making the affinity of the hydrophobic analyte for the hydrophobic stationary phase stronger relative to the now more hydrophilic mobile phase. Similarly, an investigator can decrease retention time by adding more organic solvent to the eluent. RP-HPLC is so commonly used that it is often incorrectly referred to as "HPLC" without further specification. The pharmaceutical industry regularly employs RP-HPLC to qualify drugs before their release.

RP-HPLC operates on the principle of hydrophobic interactions, which originates from the high symmetry in the dipolar water structure and plays the most important role in all processes in life science. RP-HPLC allows the measurement of these interactive forces. The binding of the analyte to the stationary phase is proportional to the contact surface area around the non-polar segment of the analyte molecule upon association with the ligand on the stationary phase. This solvophobic effect is dominated by the force of water for "cavity-reduction" around the analyte and the

C_{18}-chain versus the complex of both. The energy released in this process is proportional to the surface tension of the eluent (water: 7.3×10^{-6} J/cm², methanol: 2.2×10^{-6} J/cm²) and to the hydrophobic surface of the analyte and the ligand respectively. The retention can be decreased by adding a less polar solvent (methanol, acetonitrile) into the mobile phase to reduce the surface tension of water. Gradient elution uses this effect by automatically reducing the polarity and the surface tension of the aqueous mobile phase during the course of the analysis.

Structural properties of the analyte molecule play an important role in its retention characteristics. In general, an analyte with a larger hydrophobic surface area (C–H, C–C, and generally non-polar atomic bonds, such as S-S and others) is retained longer because it is non-interacting with the water structure. On the other hand, analytes with higher polar surface area (conferred by the presence of polar groups, such as -OH, $-NH_2$, COO^- or $-NH_3^+$ in their structure) are less retained as they are better integrated into water. Such interactions are subject to steric effects in that very large molecules may have only restricted access to the pores of the stationary phase, where the interactions with surface ligands (alkyl chains) take place. Such surface hindrance typically results in less retention.

Retention time increases with hydrophobic (non-polar) surface area. Branched chain compounds elute more rapidly than their corresponding linear isomers because the overall surface area is decreased. Similarly organic compounds with single C–C bonds elute later than those with a C=C or C–C triple bond, as the double or triple bond is shorter than a single C–C bond.

Aside from mobile phase surface tension (organizational strength in eluent structure), other mobile phase modifiers can affect analyte retention. For example, the addition of inorganic salts causes a moderate linear increase in the surface tension of aqueous solutions (ca. 1.5×10^{-7} J/cm² per Mol for NaCl, 2.5×10^{-7} J/cm² per Mol for $(NH_4)_2SO_4$), and because the entropy of the analyte-solvent interface is controlled by surface tension, the addition of salts tend to increase the retention time. This technique is used for mild separation and recovery of proteins and protection of their biological activity in protein analysis (hydrophobic interaction chromatography, HIC).

Another important factor is the mobile phase pH since it can change the hydrophobic character of the analyte. For this reason most methods use a buffering agent, such as sodium phosphate, to control the pH. Buffers serve multiple purposes: control of pH, neutralize the charge on the silica surface of the stationary phase and act as ion pairing agents to neutralize analyte charge. Ammonium formate is commonly added in mass spectrometry to improve detection of certain analytes by the formation of analyte-ammonium adducts. A volatile organic acid such as acetic acid, or most commonly formic acid, is often added to the mobile phase if mass spectrometry is used to analyze the column effluent. Trifluoroacetic acid is used infrequently in mass spectrometry applications due to its persistence in the detector and solvent delivery system, but can be effective in improving retention of analytes such as carboxylic acids in applications utilizing other detectors, as it is a fairly strong organic acid. The effects of acids and buffers vary by application but generally improve chromatographic resolution.

Reversed phase columns are quite difficult to damage compared with normal silica columns; however, many reversed phase columns consist of alkyl derivatized silica particles and should never be used with aqueous bases as these will destroy the underlying silica particle. They can be used with aqueous acid, but the column should not be exposed to the acid for too long, as it can corrode the metal parts of the HPLC equipment. RP-HPLC columns should be flushed with clean solvent after use to remove residual acids or buffers, and stored in an appropriate composition of solvent. The metal content of HPLC columns must be kept low if the best possible ability to separate substances is to be retained. A good test for the metal content of a column is to inject a sample which is a mixture of 2,2'- and 4,4'-bipyridine. Because the 2,2'-bipy can chelate the metal, the shape of the peak for the 2,2'-bipy will be distorted (tailed) when metal ions are present on the surface of the silica.

Size-exclusion Chromatography

Size-exclusion chromatography (SEC), also known as gel permeation chromatography or gel filtration chromatography, separates particles on the basis of molecular size (actually by a particle's Stokes radius). It is generally a low resolution chromatography and thus it is often reserved for the final, "polishing" step of the purification. It is also useful for determining the tertiary structure and quaternary structure of purified proteins. SEC is used primarily for the analysis of large molecules such as proteins or polymers. SEC works by trapping these smaller molecules in the pores of a particle. The larger molecules simply pass by the pores as they are too large to enter the pores. Larger molecules therefore flow through the column quicker than smaller molecules, that is, the smaller the molecule, the longer the retention time.

This technique is widely used for the molecular weight determination of polysaccharides. SEC is the official technique (suggested by European pharmacopeia) for the molecular weight comparison of different commercially available low-molecular weight heparins.

Ion-exchange Chromatography

In ion-exchange chromatography (IC), retention is based on the attraction between solute ions and charged sites bound to the stationary phase. Solute ions of the same charge as the charged sites on the column are excluded from binding, while solute ions of the opposite charge of the charged sites of the column are retained on the column. Solute ions that are retained on the column can be eluted from the column by changing the solvent conditions (e.g., increasing the ion effect of the solvent system by increasing the salt concentration of the solution, increasing the column temperature, changing the pH of the solvent, etc.)

Types of ion exchangers include polystyrene resins, cellulose and dextran ion exchangers (gels), and controlled-pore glass or porous silica. Polystyrene resins allow cross linkage which increases the stability of the chain. Higher cross linkage reduces swerving, which increases the equilibration time and ultimately improves selectivity. Cellulose and dextran ion exchangers possess larger pore sizes and low charge densities making them suitable for protein separation.

In general, ion exchangers favor the binding of ions of higher charge and smaller radius.

An increase in counter ion (with respect to the functional groups in resins) concentration reduces the retention time. A decrease in pH reduces the retention time in cation exchange while an increase in pH reduces the retention time in anion exchange. By lowering the pH of the solvent in a cation exchange column, for instance, more hydrogen ions are available to compete for positions on the anionic stationary phase, thereby eluting weakly bound cations.

This form of chromatography is widely used in the following applications: water purification, preconcentration of trace components, ligand-exchange chromatography, ion-exchange chromatography of proteins, high-pH anion-exchange chromatography of carbohydrates and oligosaccharides, and others.

Bioaffinity Chromatography

This chromatographic process relies on the property of biologically active substances to form stable, specific, and reversible complexes. The formation of these complexes involves the participation of common molecular forces such as the Van der Waals interaction, electrostatic interaction, dipole-dipole interaction, hydrophobic interaction, and the hydrogen bond. An efficient, biospecific bond is formed by a simultaneous and concerted action of several of these forces in the complementary binding sites.

Aqueous Normal-phase Chromatography

Aqueous normal-phase chromatography (ANP) is a chromatographic technique which encompasses the mobile phase region between reversed-phase chromatography (RP) and organic normal phase chromatography (ONP). This technique is used to achieve unique selectivity for hydrophilic compounds, showing normal phase elution using reversed-phase solvents.

Isocratic and Gradient Elution

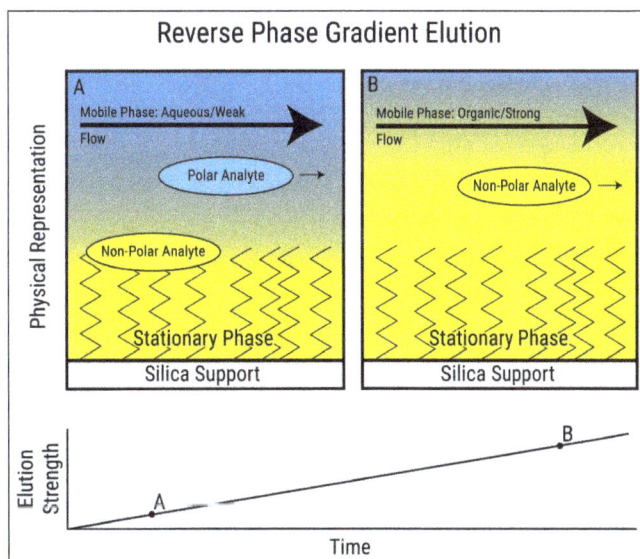

A schematic of gradient elution. Increasing mobile phase strength sequentially elutes analytes having varying interaction strength with the stationary phase.

A separation in which the mobile phase composition remains constant throughout the procedure is termed isocratic (meaning constant composition). (The example of these the percentage of methanol throughout the procedure will remain constant i.e 10%). The word was coined by Csaba Horvath who was one of the pioneers of HPLC.

The mobile phase composition does not have to remain constant. A separation in which the mobile phase composition is changed during the separation process is described as a gradient elution. One example is a gradient starting at 10% methanol and ending at 90% methanol after 20 minutes. The two components of the mobile phase are typically termed "A" and "B"; A is the "weak" solvent which allows the solute to elute only slowly, while B is the "strong" solvent which rapidly elutes the solutes from the column. In reversed-phase chromatography, solvent A is often water or an aqueous buffer, while B is an organic solvent miscible with water, such as acetonitrile, methanol, THF, or isopropanol.

In isocratic elution, peak width increases with retention time linearly according to the equation for N, the number of theoretical plates. This leads to the disadvantage that late-eluting peaks get very flat and broad. Their shape and width may keep them from being recognized as peaks.

A schematic of gradient elution. Increasing mobile phase strength sequentially elutes analytes having varying interaction strength with the stationary phase.

Gradient elution decreases the retention of the later-eluting components so that they elute faster, giving narrower (and taller) peaks for most components. This also improves the peak shape for tailed peaks, as the increasing concentration of the organic eluent pushes the tailing part of a peak forward. This also increases the peak height (the peak looks "sharper"), which is important in trace analysis. The gradient program may include sudden "step" increases in the percentage of the organic component, or different slopes at different times – all according to the desire for optimum separation in minimum time.

In isocratic elution, the selectivity does not change if the column dimensions (length and inner diameter) change – that is, the peaks elute in the same order. In gradient elution, the elution order may change as the dimensions or flow rate change.

The driving force in reversed phase chromatography originates in the high order of the water structure. The role of the organic component of the mobile phase is to reduce this high order and thus reduce the retarding strength of the aqueous component.

Parameters

Theoretical

HPLC separations have theoretical parameters and equations to describe the separation of components into signal peaks when detected by instrumentation such as by a UV detector or a mass spectrometer. The parameters are largely derived from two sets of chromatagraphic theory: plate theory (as part of Partition chromatography), and the rate theory of chromatography/Van Deemter equation. Of course, they can be put in practice through analysis of HPLC chromatograms, although rate theory is considered the more accurate theory.

They are analogous to the calculation of retention factor for a paper chromatography separation, but describes how well HPLC separates a mixture into two or more components that are detected as peaks (bands) on a chromatogram. The HPLC parameters are the: efficiency factor(N), the retention factor (kappa prime), and the separation factor (alpha). Together the factors are variables in a resolution equation, which describes how well two components' peaks separated or overlapped each other. These parameters are mostly only used for describing HPLC reversed phase and HPLC normal phase separations, since those separations tend to be more subtle than other HPLC modes (e.g., ion exchange and size exclusion).

Void volume is the amount of space in a column that is occupied by solvent. It is the space within the column that is outside of the column's internal packing material. Void volume is measured on a chromatogram as the first component peak detected, which is usually the solvent that was present in the sample mixture; ideally the sample solvent flows through the column without interacting with the column, but is still detectable as distinct from the HPLC solvent. The void volume is used as a correction factor.

Efficiency factor (N) practically measures how sharp component peaks on the chromatogram are, as ratio of the component peak's area ("retention time") relative to the width of the peaks at their widest point (at the baseline). Peaks that are tall, sharp, and relatively narrow indicate that separation method efficiently removed a component from a mixture; high efficiency. Efficiency is very dependent upon the HPLC column and the HPLC method used. Efficiency factor is synonymous with plate number, and the 'number of theoretical plates'.

Retention factor (kappa prime) measures how long a component of the mixture stuck to the column, measured by the area under the curve of its peak in a chromatogram (since HPLC chromatograms are a function of time). Each chromatogram peak will have its own retention factor (e.g., kappa1 for the retention factor of the first peak). This factor may be corrected for by the void volume of the column.

Separation factor (alpha) is a relative comparison on how well two neighboring components of the mixture were separated (i.e., two neighboring bands on a chromatogram). This factor is defined in terms of a ratio of the retention factors of a pair of neighboring chromatogram peaks, and may also be corrected for by the void volume of the column. The greater the separation factor value is over 1.0, the better the separation, until about 2.0 beyond which an HPLC method is probably not needed for separation. Resolution equations relate the three factors such that high efficiency and separation factors improve the resolution of component peaks in a HPLC separation.

Internal Diameter

Tubing on a nano-liquid chromatography (nano-LC) system, used for very low flow capacities.

The internal diameter (ID) of an HPLC column is an important parameter that influences the detection sensitivity and separation selectivity in gradient elution. It also determines the quantity of analyte that can be loaded onto the column. Larger columns are usually seen in industrial applications, such as the purification of a drug product for later use. Low-ID columns have improved sensitivity and lower solvent consumption at the expense of loading capacity.

Larger ID columns (over 10 mm) are used to purify usable amounts of material because of their large loading capacity.

Analytical scale columns (4.6 mm) have been the most common type of columns, though smaller columns are rapidly gaining in popularity. They are used in traditional quantitative analysis of samples and often use a UV-Vis absorbance detector.

Narrow-bore columns (1–2 mm) are used for applications when more sensitivity is desired either with special UV-vis detectors, fluorescence detection or with other detection methods like liquid chromatography-mass spectrometry.

Capillary columns (under 0.3 mm) are used almost exclusively with alternative detection means such as mass spectrometry. They are usually made from fused silica capillaries, rather than the stainless steel tubing that larger columns employ.

Particle Size

Most traditional HPLC is performed with the stationary phase attached to the outside of small spherical silica particles (very small beads). These particles come in a variety of sizes with 5 μm beads being the most common. Smaller particles generally provide more surface area and better separations, but the pressure required for optimum linear velocity increases by the inverse of the particle diameter squared.

This means that changing to particles that are half as big, keeping the size of the column the same, will double the performance, but increase the required pressure by a factor of four. Larger particles are used in preparative HPLC (column diameters 5 cm up to >30 cm) and for non-HPLC applications such as solid-phase extraction.

Pore Size

Many stationary phases are porous to provide greater surface area. Small pores provide greater surface area while larger pore size has better kinetics, especially for larger analytes. For example, a protein which is only slightly smaller than a pore might enter the pore but does not easily leave once inside.

Pump Pressure

Pumps vary in pressure capacity, but their performance is measured on their ability to yield a consistent and reproducible volumetric flow rate. Pressure may reach as high as 60 MPa (6000 lbf/in²), or about 600 atmospheres. Modern HPLC systems have been improved to work at much higher pressures, and therefore are able to use much smaller particle sizes in the columns (<2 μm). These ultra high performance liquid chromatography" systems or UHPLCs can work at up to 120 MPa (17,405 lbf/in²), or about 1200 atmospheres. The term "UPLC" is a trademark of the Waters Corporation, but is sometimes used to refer to the more general technique of UHPLC.

Detectors

HPLC detectors fall into two main categories: universal or selective. Universal detectors typically measure a bulk property (e.g., refractive index) by measuring a difference of a physical property between the mobile phase and mobile phase with solute while selective detectors measure a solute property (e.g., UV-Vis absorbance) by simply responding to the physical or chemical property of the solute. HPLC most commonly uses a UV-Vis absorbance detector, however, a wide range of other chromatography detectors can be used. A universal detector that complements UV-Vis absorbance detection is the Charged aerosol detector (CAD). A kind of commonly utilized detector includes refractive index detectors, which provide readings by measuring the changes in the refractive index of the effluent as it moves through the flow cell. In certain cases, it is possible to use multiple detectors, for example LCMS normally combines UV-Vis with a mass spectrometer.

Autosamplers

Large numbers of samples can be automatically injected onto an HPLC system, by the use of HPLC autosamplers. In addition, HPLC autosamplers have an injection volume and technique which is exactly the same for each injection, consequently they provide a high degree of injection volume precision.

Applications

Manufacturing

HPLC has many applications in both laboratory and clinical science. It is a common technique used in pharmaceutical development, as it is a dependable way to obtain and ensure product purity. While HPLC can produce extremely high quality (pure) products, it is not always the primary method used in the production of bulk drug materials. According to the European pharmacopoeia, HPLC is used in only 15.5% of syntheses. However, it plays a role in 44% of syntheses in the United States pharmacopoeia. This could possibly be due to differences in monetary and time constraints, as HPLC on a large scale can be an expensive technique. An increase in specificity, precision, and accuracy that occurs with HPLC unfortunately corresponds to an increase in cost.

Legal

This technique is also used for detection of illicit drugs in urine. The most common method of drug detection is an immunoassay. This method is much more convenient. However, convenience comes at the cost of specificity and coverage of a wide range of drugs. As HPLC is a method of determining (and possibly increasing) purity, using HPLC alone in evaluating concentrations of drugs is somewhat insufficient. With this, HPLC in this context is often performed in conjunction with mass spectrometry. Using liquid chromatography instead of gas chromatography in conjunction with MS circumvents the necessity for derivitizing with acetylating or alkylation agents, which can be a burdensome extra step. This technique has been used to detect a variety of agents like doping agents, drug metabolites, glucuronide conjugates, amphetamines, opioids, cocaine, BZDs, ketamine, LSD, cannabis, and pesticides. Performing HPLC in conjunction with Mass spectrometry reduces the absolute need for standardizing HPLC experimental runs.

Similar assays can be performed for research purposes, detecting concentrations of potential clinical candidates like anti-fungal and asthma drugs. This technique is obviously useful in observing multiple species in collected samples, as well, but requires the use of standard solutions when information about species identity is sought out. It is used as a method to confirm results of synthesis reactions, as purity is essential in this type of research. However, mass spectrometry is still the more reliable way to identify species.

Medical

Medical use of HPLC can include drug analysis, but falls more closely under the category of nutrient analysis. While urine is the most common medium for analyzing drug concentrations, blood serum is the sample collected for most medical analyses with HPLC. Other methods of detection of molecules that are useful for clinical studies have been tested against HPLC, namely immunoassays. In one example of this, competitive protein binding assays (CPBA) and HPLC were compared for sensitivity in detection of vitamin D. Useful for diagnosing vitamin D deficiencies in children, it was found that sensitivity and specificity of this CPBA reached only 40% and 60%, respectively, of the capacity of HPLC. While an expensive tool, the accuracy of HPLC is nearly unparalleled.

Applications of HPLC

HPLC (high-performance liquid chromatography) is one of the advanced types of chromatography.

It is highly sophisticated and expensive tools in the present analytical chemistry.

It is given prominent importance due to its attributes like:

- High sensitivity, i.e., the ability to evaluate samples of very minute concentrations like in nano-gram and picogram.

- Detect precisely chemically similar molecules like monoamines and also.

- Ability to identify compounds with complex chemistry.

This is possible in HPLC chromatography due to efficient separation under pressure over a large surface area. Besides, the HPLC system is also connected to highly sensitive detectors like UV-visible and fluorescence spectrometers, electrochemical detectors, etc.

This method of chromatography finds vast use in:

- Clinical diagnosis of diseases, disorders.

- In scientific research for discovery.

- In pharmaceutical labs for analysis.

- In the food industry for quality control.

- For standards control by government.

- For separation of similar molecules.

HPLC Analysis in the Clinical Diagnosis and Health Industry

Many disorders related to body metabolism, those related to endocrine and exocrine gland secretion, alteration in body fluids are diagnosed by HPLC analysis of concerned fluids.

For example estimation of metabolites of purines, pyrimidines or other metabolites from plasma, cerebrospinal fluid and urine samples in patients.

Estimation of corticoids from plasma in disorders of the adrenal gland which secretes an endocrine hormone.

Because of the time factor, most of the diagnostic methods are replaced by Elisa, electrophoresis and RIA methods. But still for a new a rare problem, the HPLC method is preferred to pinpoint the cause of disorders (i.e., any change in some biochemistry).

HPLC Applications in Scientific Research

HPLC system is a mandatory tool in most of the labs involved in research. The fields of study include medical, biological, chemical, biochemical, phytochemical (plant chemical research).

When research is taken up, the scientists are not sure of the actual which need attention in a body fluid or drug sample, etc. Then they have to screen every possible molecule to point out the altered change (component). Then HPLC is much suited as it can analyze every molecule in the mixture.

It finds its application to analyze and quantify the molecules. Components with similar chemistry and properties are easily distinguished by this method. Due to the principle of separation in HPLC similar molecules get separated and hence their detection, identification and quantification become easier.

HPLC Applications in the Pharmaceutical Industry

In the pharmaceutical industry, the qualitative type of HPLC analysis is widely used. In the research and development wing, both qualitative and quantitative methods are employed.

- In quality control, it is used to check if the manufactured products comply with the specified standards. These specific standards are fixed by the pharmacopeias and other drug regulating bodies. The guidelines mentioned in the pharmacopeia will give an idea of how the peak of the drug in the formulation should look, when run with specified HPLC mobile phases are used. If the peaks do not correspond to those shown in pharmacopeia, the batch cannot be passed for quality check.

- In R & D, it is used to identify the specific molecule or component in the mixture under research. Further, it is used for bioavailability studies, drug release from the formulation, dissolution studies, etc. After a formulation is designed, the drug release over some time is tested in bioavailability studies. Then the sample released is taken and injected into the HPLC system to note the individual molecules released in terms of quantity. Since the molecules might be similar, their separation is easier over the column under pressure. Further, their detection becomes easier as the system is connected UV-visible detector or other specified detectors.

For this, the drug formulations like injections, solutions, dissolved form of solid dosage forms are injected into HPLC injector to record the peaks of the individual constituents.

- Also, any new molecule under development or in a preclinical trial, are analyzed to see their concentration in the blood after certain intervals of administration. This helps to evaluate the metabolic profile, plasma concentration, bioavailability, etc. of the formulation or chemical moieties under development.

- In plant constituents, there are many molecules with similarity in chemistry like isoflavones, glycosides, saponins, etc. but the different activity or nutritional value. These compound can't be precisely determined by other methods. Hence they are determined by HPLC analysis through separation into individual components and thereby identification.

For standards control by governments: The pharmacopeia making bodies like United States Pharmacopeia (USP), British Pharmacopeia (B.P) and others use HPLC extensively. They fix standards of control for any drug formulation the industry makes. The companies send the formulation to the pharmacopeia bodies for standardization. Most formulations are estimated by HPLC to see the peaks of active ingredients (drug molecules). The peaks are then published in the official volumes of USP, IP or BP for reference by the industries for quality control. This gives an idea of how the peaks for the active ingredient in a formulation appear under the specified mobile phase solvents.

The effectiveness and use of HPLC application in recent days are further enhanced due to coupling with detectors like Mass Spectrometer, Nuclear Magnetic Resonance spectrometer, etc.

However, unlike other analytical techniques, HPLC analysis is time taking consuming process. A test run could run from a few hours to days together. It also requires expert troubleshooting in case of system failure, improper chromatogram peaks, etc. So one is advised to learn it from an expert in HPLC operation.

Clinical Diagnosis

Catecholamines such as epinephrine and dopamine are highly important for many biological functions. Analyzing their precursors and metabolites can provide diagnosis of diseases such as Parkinson's disease, heart disease, and muscular dystrophy.

However, given how physiologically widespread these molecules are, their analysis and subsequent conclusions about patient health must be done carefully. HPLC has the ability to separate and compare molecules to a higher magnitude than other techniques, making it a great candidate for such diagnostic purposes.

Reversed-phase HPLC (RP-HPLC) is one of the more popular methods due to its speed, column stability, and capacity to separate a wide range of compounds.

Identification of molecules in HPLC is done by measuring retention time. Retention time is the time it takes a molecule to pass through a column lined with adsorbents which interact differently with different molecules. This is done under varying conditions. In 1976, the potential use for RP-HPLC in diagnostic settings was shown.

Researchers exploited hydrophobic properties to separate catecholamine metabolites and amines in the same run, thereby speeding up the process. This is partly due to an interaction with pH, as acidic catecholamine metabolites are retained for longer at low pH values, but vice versa for amines.

Several conditions and settings can be modified in HPLC protocols. HPLC can then be used not only to detect diseases as mentioned, but also to monitor the progression of diseases.

Pheochromocytoma is a potentially fatal tumor of the sympathetic nervous system. It is derived from tissue in the neural crest, which implies that it secretes catecholamines. It can cause hypertension, which can complicate diagnosis, because it may only differ from hypertension in the format of its metabolites.

This makes HPLC ideal for diagnosis, however, the origin of the sample to be analyzed can affect the results. Urinary samples will reflect metabolites from both the central nervous system and the periphery.

Using cerebrospinal fluid offers results more localized to the central nervous system, and is therefore preferred.

Benefits of HPLC

High performance liquid chromatography (HPLC) is a method used to identify, quantify, or purify samples. The biological, pharmaceutical, food and drink, and environmental industries all use HPLC to test samples and substances.

HPLC works by separating compounds within a sample using high amounts of pressure. Because the compounds within the sample are all different, with varying weights, density, and composition, they will separate at different rates. This allows each individual compound to be tested and analyzed separately. This process can be used to purify water, investigate for the presence of drugs in blood, or even evaluate the composition of pharmaceutical drugs. Those with HPLC training have valuable skills that give them access to a wide variety of fields and industries where they can start their careers after graduation.

1. High performance liquid chromatography allows you to test a great diversity of samples: One of the great benefits of high performance liquid chromatography is that it gives companies the ability to test a seemingly endless variety of samples. From bio-molecules to ions, HPLC is able to stay highly accurate while testing a range of samples for different purposes. This can be a huge benefit for labs that sample many different substances, because it reduces the amount of technology they need to perform tests.

2. The testing process is customizable depending on the sample: The reason why HPLC can be used to test so many different samples is because of its customizability. As you will learn during your HPLC training, the separation of compounds occurs during the stationary phase. The stationary phase is contained within the column of an HPLC machine. There are four different types of stationary phases that can be used depending on the compound being tested and its qualities.

The most common type of stationary phase is reverse phase separation. However, HPLC technicians may also choose from normal phase, ion exchange, or size exclusion separation. This may sound confusing, but during your HPLC courses you will learn how to distinguish which type of stationary phase is best for testing different samples. Being able to adjust the separation method depending on the sample is one of the top benefits of using HPLC.

3. High performance liquid chromatography is highly efficient: Before HPLC became a popular testing method, a method called Thin-layer Chromatography (TLC) was used. HPLC is the welcomed successor of TLC, as it is much more efficient. TLC relies on gravity to move the sample, while HPLC uses a high pressure pump to move the sample along. HPLC is so efficient that it can accurately test a substance in about 10 to 30 minutes. In addition, the HPLC machine is mostly automated. Once the sample is in the machine, it performs the testing on its own. This makes it easier for lab technicians to complete documentation or analyze the results of other tests while waiting for the HPLC machine to finish.

References

- Liquid-chromatography, definition: chemicool.com, Retrieved 4 June, 2019

- Snyder, Lloyd R.; Dolan, John W. (2006). High-Performance Gradient Elution: The Practical Application of the Linear-Solvent-Strength Model. Wiley Interscience. ISBN 978-0470055519

- HPLC-Basics-principles-and-parameters, Analytical-HPLC-UHPLC, Systems-Solutions: knauer.net, Retrieved 1 April, 2019

- How-Does-High-Performance-Liquid-Chromatography-Work%3F, en-US, waters: waters.com, Retrieved 2 July, 2019

- Application-hplc-pharmaceutical-industry: studyread.com, Retrieved 1 April, 2019

- High-Performance-Liquid-Chromatography-(HPLC)-Applications, life-sciences: news-medical.net, Retrieved 20 June, 2019

- 3-benefits-high-performance-liquid-chromatography: aapscollege.ca, Retrieved 30 March, 2019

Paper Chromatography 4

- **Working of PC**
- **Paper Chromatography Procedure**
- **Applications of Paper Chromatography**
- **Pigments and Polarity**
- **Types or Modes of Paper Chromatography**
- **Advantages and Disadvantages of Paper Chromatography**

The analytical method which is used to separate colored chemicals and substances is known as paper chromatography. Some of its types are ascending chromatography, two-dimensional chromatography and descending chromatography. The diverse applications of paper chromatography as well as its advantages and disadvantages have been thoroughly discussed in this chapter.

Paper chromatography is one of the types of chromatography procedures which runs on a piece of specialized paper. It is a planar chromatography system wherein a cellulose filter paper acts as a stationary phase on which the separation of compounds occurs.

Principle of Paper Chromatography

The principle involved is partition chromatography wherein the substances are distributed or partitioned between liquid phases. One phase is the water, which is held in the pores of the filter paper used; and other is the mobile phase which moves over the paper. The compounds in the mixture get separated due to differences in their affinity towards water (in stationary phase) and mobile phase solvents during the movement of mobile phase under the capillary action of pores in the paper.

The principle can also be adsorption chromatography between solid and liquid phases, wherein the stationary phase is the solid surface of paper and the liquid phase is of mobile phase. But most of the applications of paper chromatography work on the principle of partition chromatography, i.e. partitioned between to liquid phases.

Working of PC

This is useful for separating complex mixtures of similar compounds, for example, amino acids.

A small, ideally concentrated spot of solution that contains the sample is applied to a strip of chromatography paper about 1 cm from the base, usually using a capillary tube for maximum precision. This sample is absorbed onto the paper and may form interactions with it. Any substance that reacts or bonds with the paper cannot be measured using Solvent front technique. The paper is then dipped into a suitable solvent, such as ethanol or water, taking care that the spot is above the surface of the solvent, and placed in a sealed container.

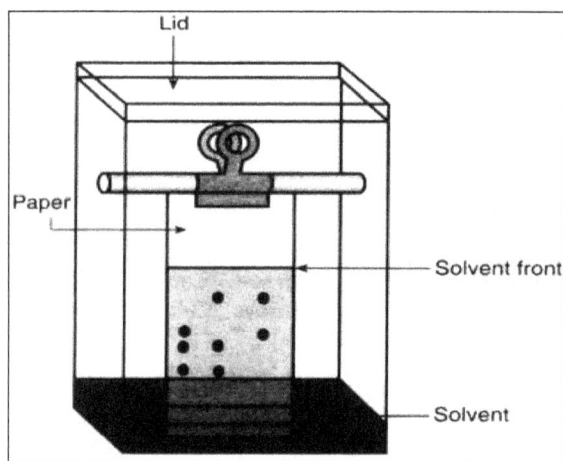

Schematic diagram of paper chromatography.

The solvent moves up the paper by capillary action, which occurs as a result of the attraction of the solvent molecules to the paper and to one another. As the solvent rises through the paper it meets and dissolves the sample mixture, which will then travel up the paper with the solvent. Different compounds in the sample mixture travel at different rates due to differences in solubility in the solvent, and due to differences in their attraction to the fibers in the paper. Paper chromatography takes anywhere from several minutes to several hours.

Paper Chromatography Procedure

The procedure to conduct Paper Chromatography Experiment are:

- Selecting a suitable type of development: It is decided based on the complexity of the solvent, paper, mixture, etc. Usually ascending type or radial paper chromatography is used as they are easy to perform. Also, it is easy to handle, the chromatogram obtained is faster and the process is less time-consuming.

- Selecting a suitable filter paper: Selection of filter paper is done based on the size of the pores, and the sample quality.

- Prepare the sample: Sample preparation includes the dissolution of the sample in a suitable solvent (inert with the sample under analysis) used in making the mobile phase.

- Spot the sample on the paper: Samples should be spotted at a proper position on the paper by using a capillary tube.

- Chromatogram development: Chromatogram development is spotted by immersing the paper in the mobile phase. Due to the capillary action of paper, the mobile phase moves over the sample on the paper.

- Paper drying and compound detection: Once the chromatogram is developed, the paper is dried using an air drier. Also, detecting solution can be sprayed on the chromatogram developed paper and dried to identify the sample chromatogram spots.

Applications of Paper Chromatography

Paper chromatography is specially used for the separation of a mixture having polar and non-polar compounds:

- For separation of amino acids.

- It is used to determine organic compounds, biochemicals in urine, etc.

- In the pharma sector it is used for the determination of hormones, drugs, etc.

- Sometimes it is used for evaluation of inorganic compounds like salts and complexes.

Paper chromatography is one of the kinds of chromatography which is used for separating complex mixtures and identifying the analytes or components. The mixture gets separated since several analytes will attract more to stationary phase (piece of specialized paper) and several components can be attracted to the mobile phase hence they travel with it. Paper chromatography is typically utilized as an instructing to explain the basic functions of chromatography, but it is used in some applications.

- Reaction monitoring.

- To ensure the control of the purity of pharmaceuticals.

- For the study of ripening and fermentation.

- For the analysis of the reaction mix in biochemical laboratories.

- To detect contaminated substances in beverages and foodstuffs.

- For the analysis of cosmetics.

- Separation and purification techniques for components.

- Forensic Testing.

- Performance enhancing drug testing.

Pigments and Polarity

Paper chromatography is one method for testing the purity of compounds and identifying substances. Paper chromatography is a useful technique because it is relatively quick and requires only small quantities of material. Separations in paper chromatography involve the same principles as those in thin layer chromatography, as it is a type of thin layer chromatography. In paper chromatography, substances are distributed between a stationary phase and a mobile phase. The stationary phase is the water trapped between the cellulose fibers of the paper. The mobile phase is a developing solution that travels up the stationary phase, carrying the samples with it. Components of the sample will separate readily according to how strongly they adsorb onto the stationary phase versus how readily they dissolve in the mobile phase.

When a colored chemical sample is placed on a filter paper, the colors separate from the sample by placing one end of the paper in a solvent. The solvent diffuses up the paper, dissolving the various molecules in the sample according to the polarities of the molecules and the solvent. If the sample contains more than one color, that means it must have more than one kind of molecule. Because of the different chemical structures of each kind of molecule, the chances are very high that each molecule will have at least a slightly different polarity, giving each molecule a different solubility in the solvent. The unequal solubility causes the various color molecules to leave solution at different places as the solvent continues to move up the paper. The more soluble a molecule is, the higher it will migrate up the paper. If a chemical is very non-polar it will not dissolve at all in a very polar solvent. This is the same for a very polar chemical and a very non-polar solvent.

It is very important to note that when using water (a very polar substance) as a solvent, the more polar the color, the higher it will rise on the papers.

Types or Modes of Paper Chromatography

Based on the way the development of chromatogram on paper is done in procedures, we have, broadly, five types of chromatography:

- Ascending chromatography: As the name indicates, the chromatogram ascends. Here, the development of paper occurs due the solvent movement or upward travel on the paper. The solvent reservoir is at the bottom of the beaker. The paper tip with sample spots just dips into the solvent at the bottom so that spots remain well above the solvent.

- Descending chromatography: Here the development of paper occurs due to solvent travel downwards on the paper. The solvent reservoir is at the top. The movement of solvent is assisted by gravity besides the capillary action.

- Ascending-descending mode: Here solvent first travels upwards and then downwards on the paper.

- Radial mode: Here the solvent travels from center (mid-point) towards the periphery of circular chromatography paper. The entire system is kept in a covered petridish for the development of the chromatogram. The wick at the center of paper dips into mobile phase in a petri dish, by which the solvent drains on to the paper and moves the sample radially to form the sample spots of different compounds as concentric rings.

- Two dimensional chromatography: Here the chromatogram development occurs in two directions at right angles.

In this mode, the samples are spotted to one corner of rectangular paper and allowed for first development. Then the paper is again immersed in mobile phase at a right angle to the previous development for second chromatogram.

Advantages and Disadvantages of Paper Chromatography

Paper chromatography is the oldest technique of separation of analytes from the sample; it is applied particularly in the analysis of lipid samples and chemical compounds. In paper chromatography, a small dot of sample applied to the piece of paper using capillary, the edge of the paper is deep in a solvent and it runs the solvent up through the paper by the capillary action. Separation of sample compounds can depend on the structure, molecular size, weight, and polarity of the compound.

Advantages of Paper Chromatography

- It requires fever quantitative material.

- Separation of compounds in a short time.

- An analysis requires a low amount of sample.

- Compare to other chromatography methods paper chromatography is a cheap technique.

- Organic as well as inorganic compounds can be identified by paper chromatography method.

- Setup of paper chromatography occupies less space than the other chromatographic method.

- Easy to handle and setup.

- The less sample quantity required for the analysis.

- Cost-effective method.

Disadvantages of Paper Chromatography

- Volatile substances cannot be separated using paper chromatography techniques.

- Paper chromatography cannot be compatible with large amounts of sample.

- Quantitative analysis is not useful in paper chromatography.

- Paper chromatography cannot be separated complex mixture.

- As compared to the HPLC or HPTLC, paper chromatography has less accuracy.

- Data cannot be saved for long periods.

References

- What-is-Paper-Chromatography-Principle-Uses-experiment-video, stem: owlcation.com, Retrieved 5 August, 2019

- Top-12-types-of-chromatographic-techniques-biochemistry, chromatography-techniques, biochemistry: biologydiscussion.com, Retrieved 6 January, 2019

- Paper-chromatography, chemistry: byjus.com, Retrieved 13 May, 2019

- Applications-of-paper-chromatography: chrominfo.blogspot.com, 1 February, 2019

- Haslam, Edwin (2007). "Vegetable tannins – Lessons of a phytochemical lifetime". Phytochemistry. 68 (22–24): 2713–21. Doi:10.1016/j.phytochem.2007.09.009. PMID 18037145

- Advantages-and-disadvantages-of-paper: chrominfo.blogspot.com, Retrieved 8 January, 2019

Thin Layer Chromatography

5

- Process of TLC
- Applications of TLC
- Advantages and Disadvantages of Thin Layer Chromatography

Thin-layer chromatography is a technique in chromatography that is used for the separation of non-volatile mixtures. This chapter has been carefully written to provide an easy understanding of the basic principle as well as various processes, applications, advantages and disadvantages of thin-layer chromatography.

Thin layer chromatography (TLC) is a chromatographic technique used to separate the components of a mixture using a thin stationary phase supported by an inert backing. It may be performed on the analytical scale as a means of monitoring the progress of a reaction, or on the preparative scale to purify small amounts of a compound. TLC is an analytical tool widely used because of its simplicity, relative low cost, high sensitivity, and speed of separation. TLC functions on the same principle as all chromatography: A compound will have different affinities for the mobile and stationary phases, and this affects the speed at which it migrates. The goal of TLC is to obtain well defined, well separated spots.

Principle of TLC

Like other chromatographic techniques, thin layer chromatography (TLC) depends on the separation principle. The separation relies on the relative affinity of compounds towards both the phases. The compounds in the mobile phase move over the surface of the stationary phase. The movement occurs in such a way that the compounds which have a higher affinity to the stationary phase move slowly while the other compounds travel fast. Therefore, the separation of the mixture is attained. On completion of the separation process, the individual components from the mixture appear as spots at respective levels on the plates. Their character and nature are identified by suitable detection techniques.

Process of TLC

The process is similar to paper chromatography with the advantage of faster runs, better separations, and the choice between different stationary phases. Because of its simplicity and speed

TLC is often used for monitoring chemical reactions and for the qualitative analysis of reaction products.

To run a thin layer chromatography plate, the following procedure is carried out:

- Using a capillary, a small spot of solution containing the sample is applied to a plate, about 1.5 centimeters from the bottom edge. The solvent is allowed to completely evaporate off to prevent it from interfering with sample's interactions with the mobile phase in the next step. If a non-volatile solvent was used to apply the sample, the plate needs to be dried in a vacuum chamber. This step is often repeated to ensure there is enough analyte at the starting spot on the plate to obtain a visible result. Different samples can be placed in a row of spots the same distance from the bottom edge, each of which will move in its own adjacent lane from its own starting point.

- A small amount of an appropriate solvent (eluent) is poured into a glass beaker or any other suitable transparent container (separation chamber) to a depth of less than 1 centimeter. A strip of filter paper (aka "wick") is put into the chamber so that its bottom touches the solvent and the paper lies on the chamber wall and reaches almost to the top of the container. The container is closed with a cover glass or any other lid and is left for a few minutes to let the solvent vapors ascend the filter paper and saturate the air in the chamber. (Failure to saturate the chamber will result in poor separation and non-reproducible results).

- The TLC plate is then placed in the chamber so that the spot(s) of the sample do not touch the surface of the eluent in the chamber, and the lid is closed. The solvent moves up the plate by capillary action, meets the sample mixture and carries it up the plate (elutes the sample). The plate should be removed from the chamber before the solvent front reaches the top of the stationary phase (continuation of the elution will give a misleading result) and dried.

- Without delay, the solvent front, the furthest extent of solvent up the plate, is marked.

- The plate is visualized. As some plates are pre-coated with a phosphor such as zinc sulfide, allowing many compounds to be visualized by using ultraviolet light; dark spots appear where the compounds block the UV light from striking the plate. Alternatively, plates can be sprayed or immersed in chemicals after elution. Various visualising agents react with the spots to produce visible results.

Plate Preparation

TLC plates are usually commercially available, with standard particle size ranges to improve reproducibility. They are prepared by mixing the adsorbent, such as silica gel, with a small amount of inert binder like calcium sulfate (gypsum) and water. This mixture is spread as a thick slurry on an unreactive carrier sheet, usually glass, thick aluminum foil, or plastic. The resultant plate is dried and activated by heating in an oven for thirty minutes at 110 °C. The thickness of the absorbent layer is typically around 0.1 – 0.25 mm for analytical purposes and around 0.5 – 2.0 mm for preparative TLC.

Processes in the Developing Chamber

TLC and HPTLC differ from all other chromatographic techniques in the fact that in addition to stationary and mobile phases, a gas phase is present. This gas phase can significantly influence the result of the separation.

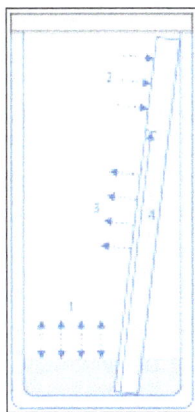

The "classical" way to develop a TLC/HPTLC plate is to place it in a chamber,
which contains a sufficient amount of developing solvent.

The lower end of the plate should be immersed several millimeters. Driven by capillary action the developing solvent moves up the layer until the desired running distance is reached and development is stopped. The following considerations primarily concern silica gel as stationary phase and developments, which can be described as adsorption chromatography.

Provided the chamber is closed, four partially competing processes occur:

- Between the components of the developing solvent and their vapor, an equilibrium will be established eventually. This equilibrium is called chamber saturation. Depending on the vapor pressure of the individual components the composition of the gas phase can differ significantly from that of the developing solvent.

- While still dry, the stationary phase adsorbs molecules from the gas phase. This process, adsorptive saturation, is also approaching an equilibrium in which the polar components will be withdrawn from the gas phase and loaded onto the surface of the stationary phase.

- Simultaneously the part of the layer which is already wetted with mobile phase interacts with the gas phase. Thereby especially the less polar components of the liquid are released into in the gas phase . Unlike this process is not as much governed by vapor pressure as by adsorption forces.

- During migration, the components of the mobile phase can be separated by the stationary phase under certain conditions, causing the formation of secondary fronts.

In connection with the development process, the following aspects should be considered.

With the exception of single component liquids (neat solvents), developing solvent and mobile phase are, strictly speaking, not the same. Their composition changes during development. Unfortunately, the terms "developing solvent" and "mobile phase" are often used as synonyms. In the true sense only the liquid in the chamber should be called developing solvent, while the liquid moving through the layer constitutes the mobile phase. Only the composition of the developing solvent at the time when it is placed into the chamber is positively known. The processes (1) and (2) can be experimentally affected by:

- Fitting one side of the chamber with filter paper soaked with developing solvent.

- Waiting a certain time between the introduction of developing solvent into the chamber and the beginning of chromatography – chamber saturation.

- Allowing the plate to interact with the gas phase prior to chromatographic development, i.e. without contact to the developing solvent – preconditioning.

An interaction according to and can be effectively prevented by placing a counter plate at a distance of one or a few millimeters to the chromatographic layer. This is called "sandwich configuration". The further an equilibrium according to and/or has been established and the less different the components of the mobile phase are in respect to their adsorption behavior, the less pronounced is the formation of secondary fronts resulting from. In well-saturated chambers and on preconditioned layers secondary fronts are often not observed. In sandwich configuration and particularly in OPLC secondary fronts are very prominent.

During chromatography, components of the developing solvent, which have been loaded onto the dry layer via the gas phase according to , are pushed ahead of the true but invisible solvent front. Exceptions are very polar components such as water, methanol, acids, or bases. This results in RF values being lower in saturated chambers and particularly on pre-conditioned layers, than in unsaturated chambers and sandwich configurations.

Note that due to possible demixing of the solvents and possible β-fronts, development in sandwich configuration or in an unsaturated horizontal developing chamber works best with single component solvents or multi component solvents behaving like single component solvents (azeotropic mixtures).

Plate and Chamber Formats

These format definitions are used in all CAMAG literature.

Some plates can be developed in one direction only, e.g. plates with a concentration zone, GLP coded plates, etc. When you order plates make sure you understand the manufacturer's size definitions.

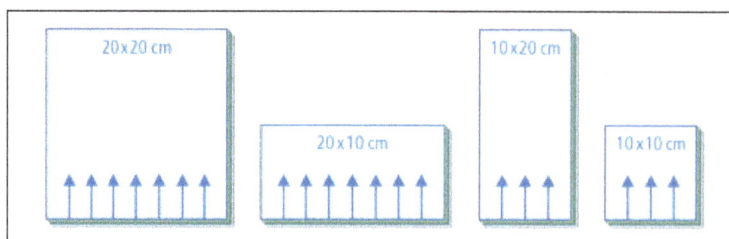

Consequences

Thin-layer Chromatography in most cases proceeds in a non-equilibrium between stationary, mobile, and gas phase. For this reason it is very difficult to correctly describe the conditions in a developing chamber.

Reproducible chromatographic results can only be expected when all parameters are kept as constant as possible. Chamber shape and saturation are playing a predominant role in this regard. Unfortunately, this means that the chromatographic result is different in each chamber.

There are neither "good" nor "poor" chambers! However, in some chambers the parameters can be better controlled, i.e. reproduced, than in others. According to the general chapters < 203 > (USP) and 2.8.28 (Ph.Eur.) a saturated chamber with the dimension 20 x 10 cm is recommended for standardized HPTLC.

Choosing a Developing Chamber

Selection of the "proper" chamber is done during method development and generally follows "practical" considerations such as which chamber is available, which one must be used due to an SOP or following a guideline, or which one has been used in the past if a results comparison is to be made. However, a focus should also be on economical aspects such as time requirement and solvent consumption and on reproducible conditions.

The Horizontal Developing Chambers have proven to be exceptionally economical, flexible and reproducible in operation. Although designed for applications where the plate is developed from two sides, they are also suitable for single-sided developments in unsaturated, saturated and sandwich configuration as well as for preconditioning of HPTLC plates. Results are not comparable to vertical developing chambers.

The Automatic Developing Chamber 2 (ADC 2) is a software-controlled device for reproducible plate development. This instrument does not only eliminate any effects of the operator when introducing the plate into a saturated chamber, but also the activity of the layer prior to start of chromatography can be set and drying of the developed plate is rapid, homogeneous, and complete. For development a conventional 20 x 10 cm Twin Trough Chamber is used. This way chamber geometry and chromatographic conditions of already existing analytical procedures can be retained, but environmental and operational effects are standardized.

In case the sample contains polar and non-polar components, which must be separated in the same analysis, the principle of Automated Multiple Development (AMD 2) can be employed. Development is performed on the basis of a solvent gradient from polar to non-polar over several steps with intermediate drying.

Examples:

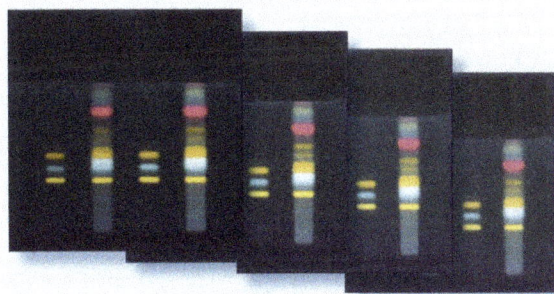

Reproducible development, here several plates under UV 366 nm.

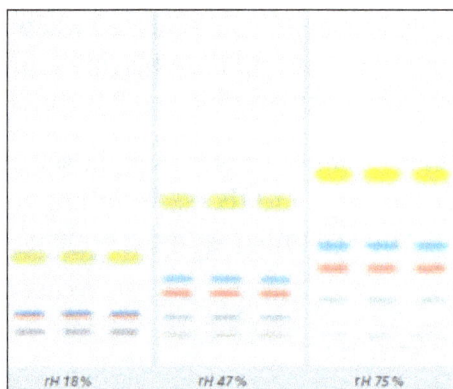

Influence of the activity of the layer (relative humidity) on the separation of a test dye mixture with toluene at equal migration distance.

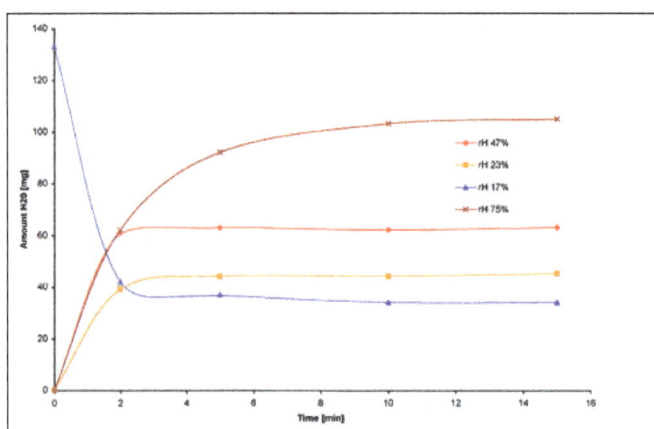

The desired activity is reached within a few minutes.

Applications of TLC

Purity of a Sample

TLC can be used to check the purity of a sample where analysis of a compound can be performed alongside an authentic reference. Presence of an impurity can be detected by the presence of extra spots on the plate.

Biochemical Analysis

TLC is often used to isolate, compare, and characterise the compounds and metabolites from blood, serum, body fluids, and urine.

Food and Cosmetic Industry

TLC can be used to separate, identify, and characterize different components colors, cosmetic products, and sweetening and preservative agents among others. It is easier to perform as it does not require any sophisticated equipment and is also time-efficient.

Pharmaceutical Industry

1. Stability Studies: Stability studies are required by drug administration where the stability of stored drugs needs to be tested. TLC can be especially useful in such tests, as several spots can be simultaneously run to compare and contrast. The results can also be easily visualised and spots which do not move or any change in the number of spots can indicate changes in the chemical nature and stability of a drug.

2. Identifying Drugs in Body Fluids: TLC can be used in forensic studies where body fluids, such as urine and blood can be tested for the presence of drugs. Acidic and neutral drugs can be identified by using octadecyl silica in the stationary phase, while plain silica and octadecyl silica can be used to identify basic drugs.

3. Identifying Drug Residues in Food: TLC can be used to identify the presence of drug residues and antibiotics in food, such as poultry, beef, pork, milk, fish among others.

4. Metabolic Profile of Drugs: Different metabolites present in a drug can be separated using TLC.

5. Monitoring the Purity of Drugs: TLC has been used to quantitatively monitor the purity of several drugs, including sedatives, antihistamines, analgesics, tranquilizers, and steroids.

Examining Reaction Mixtures

TLC can be used to study if a reaction is complete. At the beginning of a reaction, the entire spot is occupied by the starting material. During the progress of the reaction, the spot of the starting material keeps reducing, while the spot of the products keeps increasing. At the end of the reaction, the spot of the starting material is completely gone, and only one spot of the product can be seen on the plate. Thus, complete absence of the starting spot can be used to mark the end of a reaction.

Applications of thin layer chromatography in pharmaceutical analysis contain purity testing, identification, stability testing, assay, and content uniformity testing of intermediates, raw materials, and drug products, with the analysis of sample analytes. TLC is dependent on its unique detection, the speed of analysis, flexibility, and simplicity methods on together quantitative and qualitative basis. Often degradation products, synthetic intermediates, and process related impurity they do not have chromophores which cannot be detected by the UV detector. Thus, these types of impurities are frequently specified by the TLC analysis. Sometimes impurities are eluted on the solvent front in the HPLC, and they may be complicated to quantities and monitor, and modify in the mobile phase or HPLC column could not be able to solve them adequately. On the other hand, occasionally the impurities have taken more time to elute from the column, and they can't be detected, but in TLC method is open in which whole samples are evaluated. Thin layer chromatography use in the early stage of drug development.

Characterization

In organic chemistry, reactions are qualitatively monitored with TLC. Spots sampled with a capillary tube are placed on the plate: a spot of starting material, a spot from the reaction mixture, and a cross-spot with both. A small (3 by 7 cm) TLC plate takes a couple of minutes to run. The analysis is qualitative, and it will show if the starting material has disappeared, i.e. the reaction is complete, if any product has appeared, and how many products are generated (although this might be underestimated due to co-elution). Unfortunately, TLCs from low-temperature reactions may give misleading results, because the sample is warmed to room temperature in the capillary, which can alter the reaction—the warmed sample analyzed by TLC is not the same as what is in the low-temperature flask. One such reaction is the DIBALH reduction of ester to aldehyde.

In one study TLC has been applied in the screening of organic reactions, for example in the fine-tuning of BINAP synthesis from 2-naphthol. In this method, the alcohol and catalyst solution (for instance iron(III) chloride) are placed separately on the base line, then reacted, and then instantly analyzed.

A special application of TLC is in the characterization of radiolabeled compounds, where it is used to determine radiochemical purity. The TLC sheet is visualized using a sheet of photographic film or an instrument capable of measuring radioactivity. It may be visualized using other means as well. This method is much more sensitive than the others and can be used to detect an extremely small amount of a compound, provided that it carries a radioactive atom.

Isolation

| Step 1 | Step 2 | Step 3 | Step 4 | Step 5 |

| Step 6 | Step 7 |

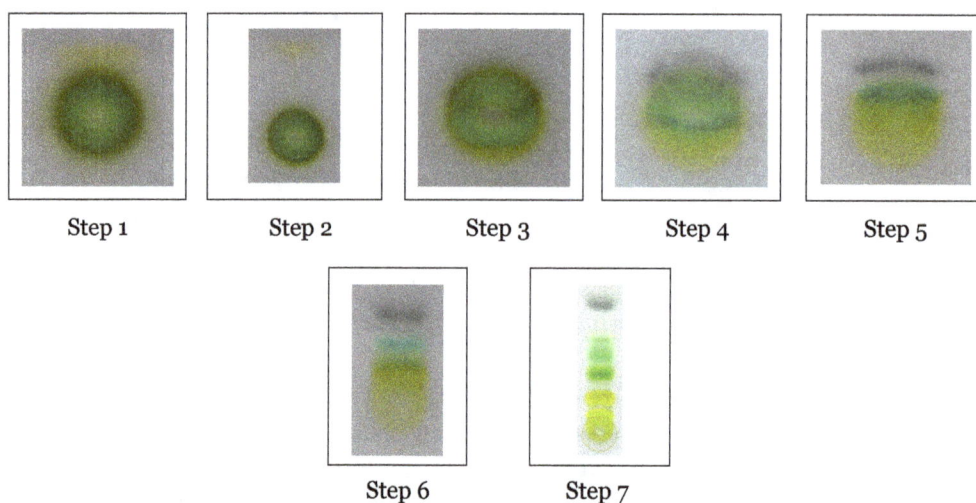

Since different compounds will travel a different distance in the stationary phase, chromatography can be used to isolate components of a mixture for further analysis. The separated compounds each occupying a specific area on the plate, they can be scraped off (along with the stationary phase particles) and dissolved into an appropriate solvent. As an example, in the chromatography of an extract of green plant material (for example spinach) shown in 7 stages of development, Carotene elutes quickly and is only visible until step 2. Chlorophyll A and B are halfway in the final step and lutein the first compound staining yellow. Once the chromatography is over, the carotene can be removed from the plate, extracted into a solvent and placed into a spectrophotometer to determine its spectrum. The quantities extracted are small and a technique such as column chromatography is preferred to separate larger amounts.

Examining Reactions

TLC is also used for the identification of the completion of any chemical reaction. To determine this it is observed that at the beginning of a reaction the entire spot is occupied by the starting chemicals or materials on the plate. As the reaction starts taking place the spot formed by the initial chemicals starts reducing and eventually replaces the whole spot of starting chemicals with a new product present on the plate. The formation of an entirely new spot determines the completion of a reaction.

Advantages and Disadvantages of Thin Layer Chromatography

Thin Layer Chromatography (TLC) is an analytical technique of separation which is used to qualitative analysis and observes the reaction of complex mixtures of analytes and also for identifying the unknown compounds. It is also significant to determine the right solvent system with which to implement in the column chromatography.

Advantages of Thin Layer Chromatography

- This is a very easy way to separate the components.

- TLC is a sensitive method.

- In comparison to other separation techniques, very few types of equipment are used. The components are separated in a very short time because the components will elute rapidly.

- It is feasible to visualize all components of UV light.

- The non-volatile compounds are separated by the TCL method.

- The only small sample size is required in TLC, and it can be in microlitre.

- A comparison with standard material, tentative identification is possible.

- The components there in the complex mixture of samples are able to easily separate and recovered by scratching the plate.

Disadvantages of Thin Layer Chromatography

- There is a no longer stationary are available in TCL plates. Therefore, its separation length is insufficient in comparison to other chromatographic techniques.

- Results obtained from TLC are difficult to reproduce.

- Only soluble components of the mixtures are possible.

- This is the only qualitative analysis possible, not quantitative.

- Usually, it is not automatic.

- TCL works in the open system, therefore, temperature and humidity can affect the results.

References

- Thin-Layer-Chromatography, General-Lab-Techniques, Demos%2C-Techniques%2C-and-Experiments, Ancillary-Materials: libretexts.org, Retrieved 21 May, 2019

- Thin-layer-chromatography, chemistry: byjus.com, Retrieved 26 July, 2019

- Chromatogram-development, products, tlc-hptlc: camag.com, Retrieved 25 February, 2019

- Applications-of-Thin-Layer-Chromatography, life-sciences: news-medical.net, Retrieved 5 May, 2019

- Applications-of-thin-layer: chrominfo.blogspot.com, Retrieved 19 April, 2019

- TLC plates as a convenient platform for solvent-free reactions Jonathan M. Stoddard, Lien Nguyen, Hector Mata-Chavez and Kelly Nguyen Chem. Commun., 2007, 1240 - 1241, doi:10.1039/b616311d

- Advantages-and-disadvantages-of-thin: chrominfo.blogspot.com, Retrieved 18 February, 2019

Techniques in Chromatography | 6

- **Reversed-phase Chromatography**

- **Simulated Moving Bed**

- **Fast protein Liquid Chromatography**

- **Countercurrent Chromatography**

- **Periodic Counter-current Chromatography**

- **Ion Chromatography**

- **Electrochromatography**

- **Capillary Electrochromatography**

- **Displacement Chromatography**

- **Size-exclusion Chromatography**

- **Micellar Electrokinetic Chromatography**

- **Two-dimensional Chromatography**

- **Affinity Chromatography**

Some of the special techniques in chromatography are reversed-phase chromatography, fast protein liquid chromatography, countercurrent chromatography, periodic counter-current chromatography, capillary electrochromatography, etc. The chapter closely examines these techniques of chromatography to provide an extensive understanding of the subject.

Chromatography is an important biophysical technique that enables the separation, identification, and purification of the components of a mixture for qualitative and quantitative analysis. Proteins can be purified based on characteristics such as size and shape, total charge, hydrophobic groups present on the surface, and binding capacity with the stationary phase. Four separation techniques based on molecular characteristics and interaction type use mechanisms of ion exchange, surface adsorption, partition, and size exclusion. Other chromatography techniques are based on the stationary bed, including column, thin layer, and paper chromatography. Column chromatography is one of the most common methods of protein purification.

Chromatography is based on the principle where molecules in mixture applied onto the surface or into the solid, and fluid stationary phase (stable phase) is separating from each other while moving with the aid of a mobile phase. The factors effective on this separation process include molecular characteristics related to adsorption (liquid-solid), partition (liquid-solid), and affinity or differences among their molecular weights. Because of these differences, some components of the mixture stay longer in the stationary phase, and they move slowly in the chromatography system, while others pass rapidly into mobile phase, and leave the system faster.

Based on this approach three components form the basis of the chromatography technique:

- Stationary phase: This phase is always composed of a "solid" phase or "a layer of a liquid adsorbed on the surface a solid support."

- Mobile phase: This phase is always composed of "liquid" or a "gaseous component."

- Separated molecules.

The type of interaction between stationary phase, mobile phase, and substances contained in the mixture is the basic component effective on separation of molecules from each other. Chromatography methods based on partition are very effective on separation, and identification of small molecules as amino acids, carbohydrates, and fatty acids. However, affinity chromatographies (ie. ion-exchange chromatography) are more effective in the separation of macromolecules as nucleic acids, and proteins. Paper chromatography is used in the separation of proteins, and in studies related to protein synthesis; gas-liquid chromatography is utilized in the separation of alcohol, esther, lipid, and amino groups, and observation of enzymatic interactions, while molecular-sieve chromatography is employed especially for the determination of molecular weights of proteins. Agarose-gel chromatography is used for the purification of RNA, DNA particles, and viruses.

Stationary phase in chromatography, is a solid phase or a liquid phase coated on the surface of a solid phase. Mobile phase flowing over the stationary phase is a gaseous or liquid phase. If mobile phase is liquid it is termed as liquid chromatography (LC), and if it is gas then it is called gas chromatography (GC). Gas chromatography is applied for gases, and mixtures of volatile liquids, and solid material. Liquid chromatography is used especially for thermal unstable, and non-volatile samples.

The purpose of applying chromatography which is used as a method of quantitative analysis apart from its separation, is to achive a satisfactory separation within a suitable timeinterval. Various chromatography methods have been developed to that end. Some of them include column chromatography, thin-layer chromatography (TLC), paper chromatography, gas chromatography, ion exchange chromatography, gel permeation chromatography, high-pressure liquid chromatography, and affinity chromatography.

Types of Chromatography

- Column chromatography;

- Ion-exchange chromatography;

- Gel-permeation (molecular sieve) chromatography;

- Affinity chromatography;

- Paper chromatography;

- Thin-layer chromatography;

- Gas chromatography;

- Dye-ligand chromatography;

- Hydrophobic interaction chromatography;

- Pseudoaffinity chromatography;

- High-pressure liquid chromatography (HPLC).

Column Chromatography

Since proteins have difference characteristic features as size, shape, net charge, stationary phase used, and binding capacity, each one of these characteristic components can be purified using chromatographic methods. Among these methods, most frequently column chromatography is applied. This technique is used for the purification of biomolecules. On a column (stationary phase) firstly the sample to be separated, then wash buffer (mobile phase) are applied. Their flow through inside column material placed on a fiberglass support is ensured. The samples are accumulated at the bottom of the device in a tme-, and volume-dependent manner.

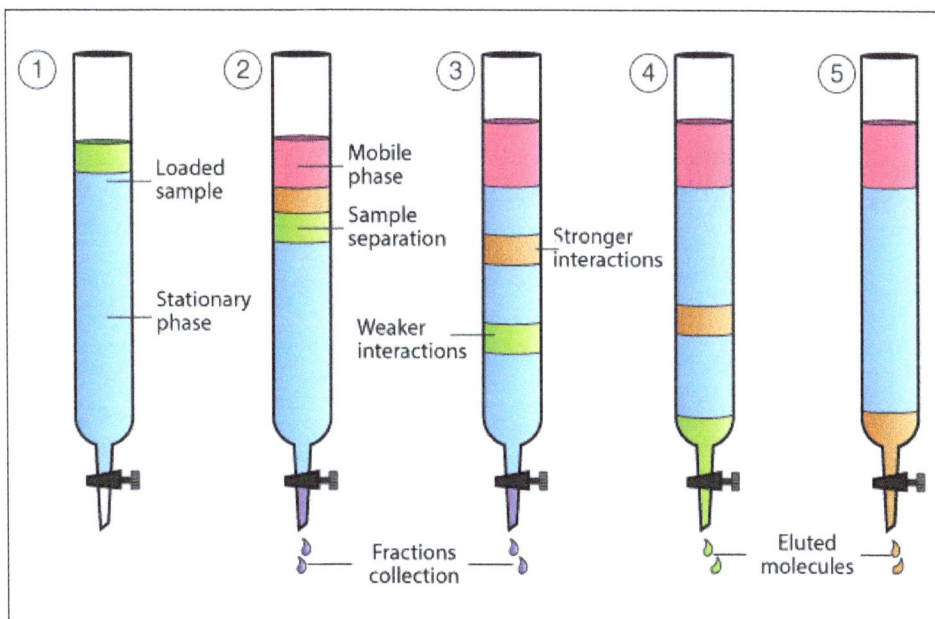

Column chromatography.

Ion-exchange Chromatography

Ion-exchange chromatography is based on electrostatic interactions between charged protein groups, and solid support material (matrix). Matrix has an ion load opposite to that of the protein to be separated, and the affinity of the protein to the column is achieved with ionic ties. Proteins are separated from the column either by changing pH, concentration of ion salts or ionic strength of the buffer solution. Positively charged ion-exchange matrices are called anion-exchange matrices, and adsorb negatively charged proteins. While matrices bound with negatively charged groups are known as cation-exchange matrices, and adsorb positively charged proteins.

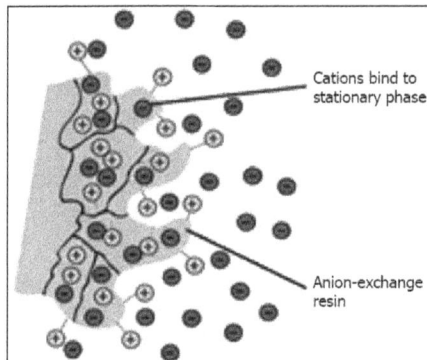

Ion-exchange chromatography.

Gel-permeation (Molecular Sieve) Chromatography

The basic principle of this method is to use dextran containing materials to separate macromolecules based on their differences in molecular sizes. This procedure is basically used to determine molecular weights of proteins, and to decrease salt concentrations of protein solutions. In a gel-permeation column stationary phase consists of inert molecules with small pores. The solution containing molecules of different dimensions are passed continuously with a constant flow rate through the column. Molecules larger than pores can not permeate into gel particles, and they are retained between particles within a restricted area. Larger molecules pass through spaces between porous particles, and move rapidly through inside the column. Molecules smaller than the pores are diffused into pores, and as molecules get smaller, they leave the column with proportionally longer retention times. Sephadeks G type is the most frequently used column material. Besides, dextran, agorose, polyacrylamide are also used as column materials.

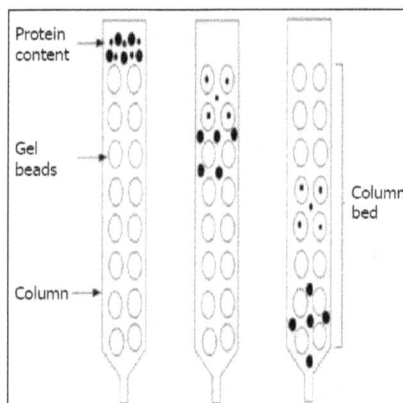

Gel-permeation (molecular sieve) chromatography.

Affinity Chromatography

This chromatography technique is used for the purification of enzymes, hormones, antibodies, nucleic acids, and specific proteins. A ligand which can make a complex with specific protein (dextran, polyacrylamide, cellulose etc) binds the filling material of the column. The specific protein which makes a complex with the ligand is attached to the solid support (matrix), and retained in the column, while free proteins leave the column. Then the bound protein leaves the column by means of changing its ionic strength through alteration of pH or addition of a salt solution.

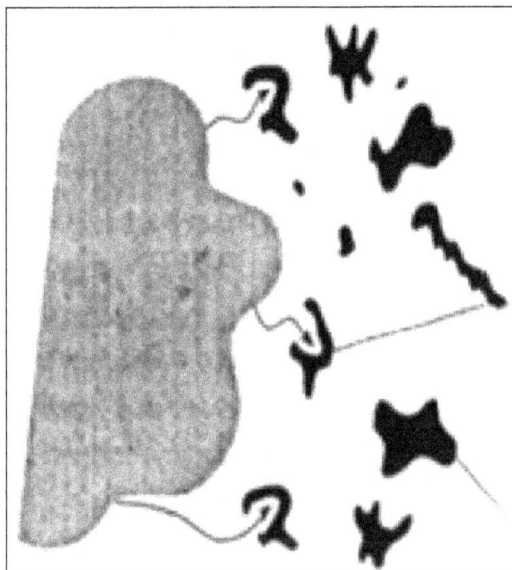

Affinity chromatography.

Paper Chromatography

In paper chromatography support material consists of a layer of cellulose highly saturated with water. In this method a thick filter paper comprised the support, and water drops settled in its pores made up the stationary "liquid phase." Mobile phase consists of an appropriate fluid placed in a developing tank. Paper chromatography is a "liquid-liquid" chromatography.

Thin-layer Chromatography

Thin-layer chromatography is a "solid-liquid adsorption" chromatography. In this method stationary phase is a solid adsorbent substance coated on glass plates. As adsorbent material all solid substances used. in column chromatography (alumina, silica gel, cellulose) can be utilized. In this method, the mobile phase travels upward through the stationary phase The solvent travels up the thin plate soaked with the solvent by means of capillary action. During this procedure, it also drives the mixture priorly dropped on the lower parts of the plate with a pipette upwards with different flow rates. Thus the separation of analytes is achieved. This upward travelling rate depends on the polarity of the material, solid phase, and of the solvent.

In cases where molecules of the sample are colorless, florescence, radioactivity or a specific chemical substance can be used to produce a visible coloured reactive product so as to identify their positions on the chromatogram. Formation of a visible colour can be observed under room light or UV light. The position of each molecule in the mixture can be measured by calculating the ratio between the the distances travelled by the molecule and the solvent. This measurement value is called relative mobility, and expressed with a symbol Rf. Rf. value is used for qualitative description of the molecules.

Gas Chromatography

In this method stationary phase is a column which is placed in the device, and contains a liquid stationary phase which is adsorbed onto the surface of an inert solid. Gas chromatography is a "gas-liquid" chromatography. Its carrier phase consists of gases as He or N_2. Mobile phase which is an inert gas is passed through a column under high pressure. The sample to be analyzed is vaporized, and enters into a gaseous mobile phase phase. The components contained in the sample are dispersed between mobile phase, and stationary phase on the solid support. Gas chromatography is a simple, multifaceted, highly sensitive, and rapidly applied technique for the extremely excellent separation of very minute molecules. It is used in the separation of very little amounts of analytes.

Dye-ligand Chromatography

Development of this technique was based on the demonstration of the ability of many enzymes to bind purine nucleotides for Cibacron Blue F3GA dye. The planar ring structure with negatively charged groups is analogous to the structure of NAD. This analogy has been evidenced by demonstration of the binding of Cibacron Blue F3GA dye to adenine, ribose binding sites of NAD. The dye behaves as an analogue of ADP-ribose. The binding capacity of this type adsorbents is 10–20-fold stronger rhat that of the affinity of other adsorbents. Under appropriate pH conditions, elution with high-ionic strength solutions, and using ion-exchange property of adsorbent, the adsorbed proteins are separated from the column.

Hydrophobic Interaction Chromatography (HIC)

In this method the adsorbents prepared as column material for the ligand binding in affinity chromatography are used. HIC technique is based on hydrophobic interactions between side chains bound to chromatography matrix.

Pseudoaffinity Chromatography

Some compounds as anthraquinone dyes, and azo-dyes can be used as ligands because of their affinity especially for dehydrogenases, kinases, transferases, and reductases The mostly known type of this kind of chromatography is immobilized metal affinity chromatography (IMAC).

Reversed-phase Chromatography

Reversed-phase chromatography (RPC) is any liquid chromatography procedure in which the mobile phase is significantly more polar than the stationary phase. It is so named because in normal-phase liquid chromatography, the mobile phase is significantly less polar than the stationary phase. Hydrophobic molecules in the mobile phase tend to adsorb to the relatively hydrophobic stationary phase. Hydrophilic molecules in the mobile phase will tend to elute first. Separating columns typically comprise a C8 or C18 carbon-chain bonded to a silica particle substrate.

Hydrophobic Interaction Chromatography

Hydrophobic interactions between proteins and the chromatographic matrix can be exploited to purify proteins. In hydrophobic interaction chromatography the matrix material is lightly substituted with hydrophobic groups. These groups can range from methyl, ethyl, propyl, octyl, or phenyl groups. At high salt concentrations, non-polar sidechains on the surface on proteins "interact" with the hydrophobic groups; that is, both types of groups are excluded by the polar solvent (hydrophobic effects are augmented by increased ionic strength). Thus, the sample is applied to the column in a buffer which is highly polar. The eluant is typically an aqueous buffer with decreasing salt concentrations, increasing concentrations of detergent (which disrupts hydrophobic interactions), or changes in pH.

In general, Hydrophobic Interaction Chromatography (HIC) is advantageous if the sample is sensitive to pH change or harsh solvents typically used in other types of chromatography but not high salt concentrations. Commonly, it is the amount of salt in the buffer which is varied. In 2012, Müller and Franzreb described the effects of temperature on HIC using Bovine Serum Albumin (BSA) with four different types of hydrophobic resin. The study altered temperature as to effect the binding affinity of BSA onto the matrix. It was concluded that cycling temperature from 50 degrees to 10 degrees would not be adequate to effectively wash all BSA from the matrix but could be very effective if the column would only be used a few times. Using temperature to effect change allows labs to cut costs on buying salt and saves money.

If high salt concentrations along with temperature fluctuations want to be avoided you can use a more hydrophobic to compete with your sample to elute it. This so-called salt independent method of HIC showed a direct isolation of Human Immunoglobulin G (IgG) from serum with satisfactory yield and used Beta-cyclodextrin as a competitor to displace IgG from the matrix. This largely opens up the possibility of using HIC with samples which are salt sensitive as we know high salt concentrations precipitate proteins.

Two-dimensional Chromatography

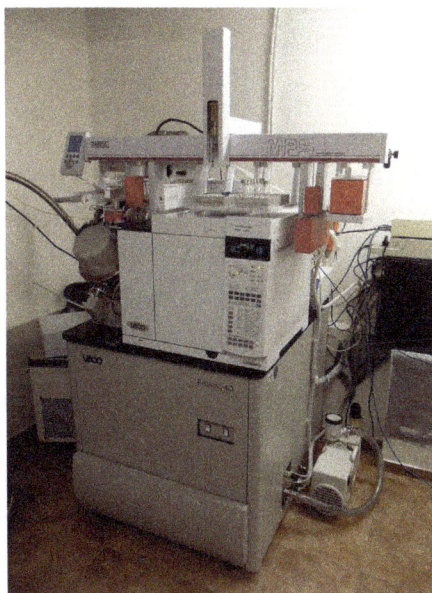

Two-dimensional chromatograph GCxGC-TOFMS.

In some cases, the chemistry within a given column can be insufficient to separate some analytes. It is possible to direct a series of unresolved peaks onto a second column with different physico-chemical (chemical classification) properties. Since the mechanism of retention on this new solid support is different from the first dimensional separation, it can be possible to separate compounds by two-dimensional chromatography that are indistinguishable by one-dimensional chromatography.

The sample is spotted at one corner of a square plate, developed, air-dried, then rotated by 90° and usually redeveloped in a second solvent system.

Simulated Moving-bed Chromatography

The simulated moving bed (SMB) technique is a variant of high performance liquid chromatography; it is used to separate particles and/or chemical compounds that would be difficult or impossible to resolve otherwise. This increased separation is brought about by a valve-and-column arrangement that is used to lengthen the stationary phase indefinitely. In the moving bed technique of preparative chromatography the feed entry and the analyte recovery are simultaneous and continuous, but because of practical difficulties with a continuously moving bed, simulated moving bed technique was proposed. In the simulated moving bed technique instead of moving the bed, the sample inlet and the analyte exit positions are moved continuously, giving the impression of a moving bed. True moving bed chromatography (TMBC) is only a theoretical concept. Its simulation, SMBC is achieved by the use of a multiplicity of columns in series and a complex valve arrangement, which provides for sample and solvent feed, and also analyte and waste takeoff at appropriate locations of any column, whereby it allows switching at regular intervals the sample entry in one direction, the solvent entry in the opposite direction, whilst changing the analyte and waste takeoff positions appropriately as well.

Pyrolysis Gas Chromatography

Pyrolysis–gas chromatography–mass spectrometry is a method of chemical analysis in which the sample is heated to decomposition to produce smaller molecules that are separated by gas chromatography and detected using mass spectrometry.

Pyrolysis is the thermal decomposition of materials in an inert atmosphere or a vacuum. The sample is put into direct contact with a platinum wire, or placed in a quartz sample tube, and rapidly heated to 600–1000 °C. Depending on the application even higher temperatures are used. Three different heating techniques are used in actual pyrolyzers: Isothermal furnace, inductive heating (Curie Point filament), and resistive heating using platinum filaments. Large molecules cleave at their weakest points and produce smaller, more volatile fragments. These fragments can be separated by gas chromatography. Pyrolysis GC chromatograms are typically complex because a wide range of different decomposition products is formed. The data can either be used as fingerprint to prove material identity or the GC/MS data is used to identify individual fragments to obtain structural information. To increase the volatility of polar fragments, various methylating reagents can be added to a sample before pyrolysis.

Besides the usage of dedicated pyrolyzers, pyrolysis GC of solid and liquid samples can be performed directly inside Programmable Temperature Vaporizer (PTV) injectors that provide quick heating (up to 30 °C/s) and high maximum temperatures of 600–650 °C. This is sufficient for some pyrolysis applications. The main advantage is that no dedicated instrument has to be purchased and pyrolysis can be performed as part of routine GC analysis. In this case quartz GC inlet liners have to be used.

Fast Protein Liquid Chromatography

Fast protein liquid chromatography (FPLC), is a form of liquid chromatography that is often used to analyze or purify mixtures of proteins. As in other forms of chromatography, separation is possible because the different components of a mixture have different affinities for two materials, a moving fluid (the "mobile phase") and a porous solid (the stationary phase). In FPLC the mobile phase is an aqueous solution, or "buffer". The buffer flow rate is controlled by a positive-displacement pump and is normally kept constant, while the composition of the buffer can be varied by drawing fluids in different proportions from two or more external reservoirs. The stationary phase is a resin composed of beads, usually of cross-linked agarose, packed into a cylindrical glass or plastic column. FPLC resins are available in a wide range of bead sizes and surface ligands depending on the application.

Countercurrent Chromatography

Countercurrent chromatography (CCC) is a type of liquid-liquid chromatography, where both the stationary and mobile phases are liquids. The operating principle of CCC equipment requires a column consisting of an open tube coiled around a bobbin. The bobbin is rotated in a double-axis gyratory motion (a cardioid), which causes a variable gravity (G) field to act on the column during each rotation. This motion causes the column to see one partitioning step per revolution and components of the sample separate in the column due to their partitioning coefficient between the two immiscible liquid phases used. There are many types of CCC available today. These include HSCCC (High Speed CCC) and HPCCC (High Performance CCC). HPCCC is the latest and best performing version of the instrumentation available currently.

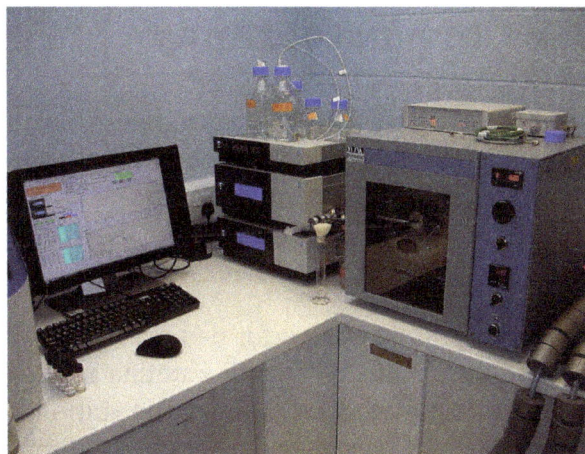

An example of a HPCCC system.

Periodic Countercurrent Chromatography

In contrast to Countercurrent chromatography, periodic counter-current chromatography (PCC) uses a solid stationary phase and only a liquid mobile phase. It thus is much more similar to conventional affinity chromatography than to countercurrent chromatography. PCC uses multiple columns, which during the loading phase are connected in line. This mode allows for overloading the first column in this series without losing product, which already breaks through the column before the resin is fully saturated. The breakthrough product is captured on the subsequent columns. In a next step the columns are disconnected from one another. The first column is washed and eluted, while the other columns are still being loaded. Once the (initially) first column is re-equilibrated, it is re-introduced to the loading stream, but as last column. The process then continues in a cyclic fashion.

Chiral Chromatography

Chiral chromatography involves the separation of stereoisomers. In the case of enantiomers, these have no chemical or physical differences apart from being three-dimensional mirror images. Conventional chromatography or other separation processes are incapable of separating them. To enable chiral separations to take place, either the mobile phase or the stationary phase must themselves be made chiral, giving differing affinities between the analytes. Chiral chromatography HPLC columns (with a chiral stationary phase) in both normal and reversed phase are commercially available.

Aqueous Normal-phase Chromatography

Aqueous normal-phase (ANP) chromatography is characterized by the elution behavior of classical normal phase mode (i.e. where the mobile phase is significantly less polar than the stationary phase) in which water is one of the mobile phase solvent system components. It is distinguished from hydrophilic interaction liquid chromatography (HILIC) in that the retention mechanism is due to adsorption rather than partitioning.

Reversed-phase Chromatography

Reversed-phase chromatography (also called RPC, reverse-phase chromatography, or hydrophobic chromatography) includes any chromatographic method that uses a hydrophobic stationary phase. RPC refers to liquid (rather than gas) chromatography.

Stationary Phases

In the 1970s, most liquid chromatography was performed using a solid support stationary phase (also called a column) containing unmodified silica or alumina resins. This type of technique is now referred to as normal-phase chromatography. Since the stationary phase is hydrophilic in this technique, molecules with hydrophilic properties contained within the mobile phase will have a high affinity for the stationary phase, and therefore will adsorb to the column packing. Hydrophobic molecules experience less of an affinity for the column packing, and will pass through to be eluted and detected first. Elution of the hydrophilic molecules adsorbed to the column packing requires the use of more hydrophilic or more polar solvents in the mobile phase to shift the distribution of the particles in the stationary phase towards that of the mobile phase.

Reversed-phase chromatography is a technique using alkyl chains covalently bonded to the stationary phase particles in order to create a hydrophobic stationary phase, which has a stronger affinity for hydrophobic or less polar compounds. The use of a hydrophobic stationary phase is essentially the reverse of normal phase chromatography, since the polarity of the mobile and stationary phases have been inverted – hence the term reversed-phase chromatography.

Reversed-phase chromatography employs a polar (aqueous) mobile phase. As a result, hydrophobic molecules in the polar mobile phase tend to adsorb to the hydrophobic stationary phase, and hydrophilic molecules in the mobile phase will pass through the column and are eluted first. Hydrophobic molecules can be eluted from the column by decreasing the polarity of the mobile phase using an organic (non-polar) solvent, which reduces hydrophobic interactions. The more hydrophobic the molecule, the more strongly it will bind to the stationary phase, and the higher the concentration of organic solvent that will be required to elute the molecule.

Many of the mathematical and experimental considerations used in other chromatographic methods also apply to RPC (for example, the separation resolution is dependent on the length of the column). It can be used for the separation of a wide variety of molecules. It is not typically used for separation of proteins, because the organic solvents used in RPC can denature many proteins. For this reason, normal phase chromatography is more commonly used for separation of proteins.

Today, RPC is a frequently used analytical technique. There are a variety of stationary phases available for use in RPC, allowing great flexibility in the development of separation methods.

Silica-based Stationary Phases

Any inert polar substance that achieves sufficient packing can be used for reversed-phase chromatography. The most popular column is an octadecyl carbon chain (C18)-bonded silica (USP classification L1). This is followed by C8-bonded silica (L7), pure silica (L3), cyano-bonded silica (L10) and phenyl-bonded silica (L11). Note that C18, C8 and phenyl are dedicated reversed-phase resins, while cyano columns can be used in a reversed-phase mode depending on analyte and mobile phase conditions. Not all C18 columns have identical retention properties. Surface functionalization of silica can be performed in a monomeric or a polymeric reaction with different short-chain organosilanes used in a second step to cover remaining silanol groups (end-capping). While the overall retention mechanism remains the same, subtle differences in the surface chemistries of different stationary phases will lead to changes in selectivity.

Modern columns have different polarity. PFP is pentafluorphenyl. CN is cyano. NH2 is amino. ODS is octadecyl or C18. ODCN is a mixed mode column consisting of C18 and nitrile. SCX is strong cationic exchange (used for separation of organic amines). SAX is strong anionic exchange (used for separation of carboxylic acid compounds).

Mobile Phases

Mixtures of water or aqueous buffers and organic solvents are used to elute analytes from a reversed-phase column. The solvents must be miscible with water, and the most common organic solvents used are acetonitrile, methanol, and tetrahydrofuran (THF). Other solvents can be used such as ethanol or 2-propanol (isopropyl alcohol). Elution can be performed isocratically (the water-solvent composition does not change during the separation process) or by using a solution gradient (the water-solvent composition changes during the separation process, usually by decreasing the polarity). The pH of the mobile phase can have an important role on the retention of an analyte and can change the selectivity of certain analytes.

Charged analytes can be separated on a reversed-phase column by the use of ion-pairing (also called ion-interaction). This technique is known as reversed-phase ion-pairing chromatography.

Simulated Moving Bed

In manufacturing, the simulated moving bed (SMB) process is a highly engineered process for implementing chromatographic separation. It is used to separate one chemical compound or one class of chemical compounds from one or more other chemical compounds to provide significant quantities of the purified or enriched material at a lower cost than could be obtained using simple (batch) chromatography. It cannot provide any separation or purification that cannot be done by a simple column purification. The process is rather complicated. The single advantage which it brings to a chromatographic purification is that it allows the production of large quantities of highly purified material at a dramatically reduced cost. The cost reductions come about as a result of: the use of a smaller amount of chromatographic separation media stationary phase, a continuous and high rate of production, and decreased solvent and energy requirements. This improved economic performance is brought about by a valve-and-column arrangement that is used to lengthen the stationary phase indefinitely and allow very high solute loadings to the process.

In the conventional moving bed technique of production chromatography the feed entry and the analyte recovery are simultaneous and continuous, but because of practical difficulties with a continuously moving bed, the simulated moving bed technique was proposed. In the simulated moving bed technique instead of moving the bed, the feed inlet, the solvent or eluent inlet and the desired product exit and undesired product exit positions are moved continuously, giving the impression of a moving bed, with continuous flow of solid particles and continuous flow of liquid in the opposite direction of the solid particles.

True moving bed chromatography (TMBC) is only a theoretical concept. Its simulation, SMBC, is achieved by the use of a multiplicity of columns in series and a complex valve arrangement, which provides for flow of the feed mixture and solvent, and "eluent" or "desorbent" feed at any column. The valving and piping arrangements and the predetermined control of these allow switching at regular intervals the sample entry in one direction, the solvent entry in the same direction but at a different location in the continuous loop, whilst changing the fast product and slow product take-off positions to also move in the same direction, but at different relative locations within the loop.

The advantage of the SMBC is high production rate, because a system could be near continuous, whilst its disadvantage is that it only performs one cut in mixtures. Thus, it is well-suited for separation of a binary mixture. With multiple cuts, analogous to a series of distillation columns, multiple compounds can be separated from a mixture of more than two compounds. With regard to efficiency it compares with the simple chromatography technique like continuous distillation does with batch distillation.

Construction

Specifically, an SMB system has two or more identical columns, which are connected to the mobile phase pump, and each other, by a multi-port valve. The plumbing is configured in such a way that:

- All columns will be connected in series, forming a single continuous loop;

- Typically, between each column there will be provisions for four process streams: incoming feed mixture, exiting purified fast component, exiting purified slow component, and incoming solvent or eluent;

- Each process stream (two inlets and two outlets) will proceed in the same direction after a set time (the steptime).

Advantages

SMB provides lower production cost by requiring less column volume, less chromatographic separation media ("packing" or "stationary phase"), using less solvent and less energy, and requiring far less labor.

At industrial scale an SMB chromatographic separator is operated continuously, requiring less resin and less solvent than batch chromatography. The continuous operation facilitates operation control and integration into production plants. Low eluent consumption High product concentration High productivity Continuous process This system is useful in the supercritical fluid extraction to obtain large quantity of specific product.

Drawbacks

The drawbacks of the SMB are higher investment cost compared to single column operations, a higher complexity, as well as higher maintenance costs. But these drawbacks are effectively compensated by the better yield and a much lower solvent consumption as well as a much higher productivity compared to simple batch separations.

For purifications, in particular the isolation of an intermediate single component or a fraction out of a multicomponent mixture, the SMB is not as ideally suited. Normally, a single SMB will separate only two fractions from each other, but a series or "train" of SMBs can perform multiple cuts and purify one or more products from a multi-component mixture. SMB is not readily suited for solvent gradients. Solvent gradient purification may be preferred for the purification of some biomolecules. A continuous chromatography technique to overcome the two fraction limit and to apply gradients is multicolumn countercurrent solvent gradient purification (MCSGP).

Applications

In size exclusion chromatography, where the separation process is driven by entropy, it is not possible to increase the resolution attained by a column via temperature or solvent gradients. Consequently, these separations often require SMB, to extend usable retention time differences between the molecules or particles being separated. SMB is also very useful in the pharmaceutical industry, where separation of molecules having different chirality must be done on a very large scale. For the purification of fructose, e.g. in high fructose corn syrup, or amino-acids, biological-acids, etc. on an industrial scale, simulated moving bed chromatography is used in order to improve the economics of the production.

Fast Protein Liquid Chromatography

Fast protein liquid chromatography (FPLC), is a form of liquid chromatography that is often used to analyze or purify mixtures of proteins. As in other forms of chromatography, separation is possible because the different components of a mixture have different affinities for two materials, a moving fluid (the "mobile phase") and a porous solid (the stationary phase). In FPLC the mobile phase is an aqueous solution, or "buffer". The buffer flow rate is controlled by a positive-displacement pump and is normally kept constant, while the composition of the buffer can be varied by drawing fluids in different proportions from two or more external reservoirs. The stationary phase is a resin composed of beads, usually of cross-linked agarose, packed into a cylindrical glass or plastic column. FPLC resins are available in a wide range of bead sizes and surface ligands depending on the application.

In the most common FPLC strategy, ion exchange, a resin is chosen that the protein of interest will bind to the resin by a charge interaction while in buffer A (the running buffer) but become dissociated and return to solution in buffer B (the elution buffer). A mixture containing one or more proteins of interest is dissolved in 100% buffer A and pumped into the column. The proteins of interest bind to the resin while other components are carried out in the buffer. The total flow rate of the buffer is kept constant; however, the proportion of Buffer B (the "elution" buffer) is gradually increased from 0% to 100% according to a programmed change in concentration (the "gradient"). At some point during this process each of the bound proteins dissociates and appears in the eluant. The eluant passes through two detectors which measure salt concentration (by conductivity) and protein concentration (by absorption of ultraviolet light at a wavelength of 280nm). As each protein is eluted it appears in the eluant as a "peak" in protein concentration and can be collected for further use.

FPLC was developed and marketed in Sweden by Pharmacia in 1982 and was originally called fast performance liquid chromatography to contrast it with HPLC or high-performance liquid chromatography. FPLC is generally applied only to proteins; however, because of the wide choice of resins and buffers it has broad applications. In contrast to HPLC the buffer pressure used is relatively low, typically less than 5 bar, but the flow rate is relatively high, typically 1-5 ml/min. FPLC can be readily scaled from analysis of milligrams of mixtures in columns with a total volume of 5ml or less to industrial production of kilograms of purified protein in columns with volumes of many liters. When used for analysis of mixtures the eluant is usually collected in fractions of 1-5 ml which can be further analyzed, e.g. by MALDI mass spectrometry. When used for protein purification there may be only two collection containers, one for the purified product and one for waste.

FPLC System Components

A typical laboratory FPLC consist of one or two high-precision pumps, a control unit, a column, a detection system and a fraction collector. Although it is possible to operate the system manually, the components are normally linked to a personal computer or, in older units, a microcontroller.

Pumps

The majority of systems utilize two two-cylinder piston pumps, one for each buffer, combining the output of both in a mixing chamber. Some simpler systems use a single peristaltic pump which draws both buffers from separate reservoirs through a proportioning valve and mixing chamber. In either case the system allows the fraction of each buffer entering the column to be continuously varied. The flow rate can go from a few milliliters per minute in bench-top systems to liters per minute for industrial scale purifications. The wide flow range makes it suitable both for analytical and preparative chromatography.

Injection Loop

The injection loop is a segment of tubing of known volume which is filled with the sample solution before it is injected into the column. Loop volume can range from a few microliters to 50 ml or more.

Injection Valve

The injection valve is a motorized valve which links the mixer and sample loop to the column. Typically the valve has three positions for loading the sample loop, for injecting the sample from the loop into the column, and for connecting the pumps directly to the waste line to wash them or change buffer solutions. The injection valve has a sample loading port through which the sample can be loaded into the injection loop, usually from a hypodermic syringe using a Luer-lock connection.

Column

The column is a glass or plastic cylinder packed with beads of resin and filled with buffer solution. It is normally mounted vertically with the buffer flowing downward from top to bottom. A glass frit at the bottom of the column retains the resin beads in the column while allowing the buffer and dissolved proteins to exit.

Flow Cell

The eluant from the column passes through one or more flow cells to measure the concentration of protein in the eluant (by UV light absorption at 280 nm). The conductivity cell measures the buffer conductivity, usually in millisiemens/cm, which indicates the concentration of salt in the buffer. A flow cell which measures pH of the buffer is also commonly included. Usually each flow cell is connected to a separate electronics module which provides power and amplifies the signal.

Monitor/Recorder

The flow cells are connected to a display and/or recorder. On older systems this was a simple chart recorder, on modern systems a computer with hardware interface and display is used. This permits the experimenter to identify when peaks in protein concentration occur, indicating that specific components of the mixture are being eluted.

Fraction Collector

The fraction collector is typically a rotating rack that can be filled with test tubes or similar containers. It allows samples to be collected in fixed volumes, or can be controlled to direct specific fractions detected as peaks of protein concentration, into separate containers.

Many systems include various optional components. A filter may be added between the mixer and column to minimize clogging. In large FPLC columns the sample may be loaded into the column directly using a small peristaltic pump rather than an injection loop. When the buffer contains dissolved gas, bubbles may form as pressure drops where the buffer exits the column; these bubbles create artifacts if they pass through the flow cells. This may be prevented by degassing the buffers, e.g. with a degasser, or by adding a flow restrictor downstream of the flow cells to maintain a pressure of 1-5 bar in the eluant line.

FPLC Columns

The columns used in FPLC are large [mm id] tubes that contain small [μ] particles or gel beads that are known as stationary phase. The chromatographic bed is composed by the gel beads inside the column and the sample is introduced into the injector and carried into the column by the flowing solvent. As a result of different components adhering to or diffusing through the gel, the sample mixture gets separated.

Columns used with an FPLC can separate macromolecules based on size, charge distribution (ion exchange), hydrophobicity, reverse-phase or biorecognition (as with affinity chromatography). For easy use, a wide range of pre-packed columns for techniques such as ion exchange, gel filtration (size exclusion), hydrophobic interaction, and affinity chromatography are available. FPLC differs from HPLC in that the columns used for FPLC can only be used up to maximum pressure of 3-4 MPa (435-580 psi). Thus, if the pressure of HPLC can be limited, each FPLC column may also be used in an HPLC machine.

Optimizing Protein Purification

Combinations of chromatographic methods can be used to purify a target molecule. The purpose of purifying proteins with FPLC is to deliver quantities of the target at sufficient purity in a biologically active state to suit its further use. The quality of the end product varies depending the type and amount of starting material, efficiency of separation, and selectivity of the purification resin. The ultimate goal of a given purification protocol is to deliver the required yield and purity of the target molecule in the quickest, cheapest, and safest way for acceptable results. The range of purity required can be from that required for basic analysis (SDS-PAGE or ELISA, for example), with only bulk impurities removed, to pure enough for structural analysis (NMR or X-ray crystallography), approaching >99% target molecule. Purity required can also mean pure enough that the biological activity of the target is retained. These demands can be used to determine the amount of starting material required to reach the experimental goal. If the starting material is limited and full optimization of purification protocol cannot be performed, then a safe standard protocol that requires a minimum adjustment and optimization steps are expected. This may not be optimal with respect to experimental time, yield, and economy but it will achieve the experimental goal. On the other hand, if the starting material is enough to develop more complete protocol, the amount of work to reach the separation goal depends on the available sample information and target molecule properties. Limits to development of purification protocols many times depends on the source of the substance to be purified, whether from natural sources (harvested tissues or organisms, for example), recombinant sources (such as using prokaryotic or eukaryotic vectors in their respective expression systems), or totally synthetic sources.

No chromatographic techniques provide 100% yield of active material and overall yields depend on the number of steps in the purification protocol. By optimizing each step for the intended purpose and arranging them that minimizes inter step treatments, the number of steps will be minimized.

A typical multistep purification protocol starts with a preliminary capture step which often utilizes ion exchange chromatography (IEC). The media (stationary phase) resin consists of beads, which range in size from being large (good for fast flow rates and little to no sample clarification at the expense of resolution) to small (for best possible resolution with all other factors being equal). Short and wide column geometries are amenable to high flow rates also at the expense of resolution, typically because of lateral diffusion of sample on the column. For techniques such as size exclusion chromatography to be useful, very long, thin columns and minimal sample volumes (maximum 5% of column volume) are required. Hydrophobic interaction chromatography (HIC) can also be used for first and/or intermediate steps. Selectivity in HIC is independent of running pH and descending salt gradients are used. For HIC, conditioning involves adding ammonium sulphate to the sample to match the buffer A concentration. If HIC is used before IEC, the ionic strength would have to be lowered to match that of buffer A for IEC step by dilution, dialysis or buffer exchange by gel filtration. This is why IEC is usually performed prior to HIC as the high salt elution conditions for IEC are ideal for binding to HIC resins in the next purification step. Polishing is used to achieve the final level of purification required and is commonly performed on a gel filtration column. An extra intermediate purification step can be added or optimization of the different steps is performed for improving purity. This extra step usually involves another round of IEC under completely different conditions.

Although this is an example of a common purification protocol for proteins, the buffer conditions, flow rates, and resins used to achieve final goals can be chosen to cover a broad range of target proteins. This flexibility is imperative for a functional purification system as all proteins behave differently and often deviate from predictions.

Fast protein liquid chromatography (FPLC) is a form of high-performance chromatography that takes advantage of high resolution made possible by small-diameter stationary phases. It was originally developed for proteins and features high loading capacity, biocompatible aqueous buffer systems, fast flow rates, and availability of stationary phases in most common chromatography modes (e.g., ion exchange, gel filtration, reversed phase, and affinity). The system makes reproducible separation possible by incorporating a high level of automation including autosamplers, gradient program control, and peak collection. In addition to proteins, the method is applicable to other kinds of biological samples including oligonucleotides and plasmids. The most common type of FPLC experiment is anion exchange of proteins.

Countercurrent Chromatography

Countercurrent chromatography (CCC, also counter-current chromatography) is a form of liquid–liquid chromatography that uses a liquid stationary phase that is held in place by centrifugal force and is used to separate, identify, and quantify the chemical components of a mixture. In its broadest sense, countercurrent chromatography encompasses a collection of related liquid chromatography techniques that employ two immiscible liquid phases without a solid support. The two liquid phases come in contact with each other as at least one phase is pumped through a column, a hollow tube or a series of chambers connected with channels, which contains both phases. The resulting dynamic mixing and settling action allows the components to be separated by their respective solubilities in the two phases. A wide variety of two-phase solvent systems consisting of at least two immiscible liquids may be employed to provide the proper selectivity for the desired separation.

Some types of countercurrent chromatography, such as dual flow CCC, feature a true countercurrent process where the two immiscible phases flow past each other and exit at opposite ends of the column. More often, however, one liquid acts as the stationary phase and is retained in the column while the mobile phase is pumped through it. The liquid stationary phase is held in place by gravity or by centrifugal force. An example of a gravity method is called droplet counter current chromatography (DCCC). There are two modes by which the stationary phase is retained by centrifugal force: hydrostatic and hydrodynamic. In the hydrostatic method, the column is rotated about a central axis. Hydrostatic instruments are marketed under the name centrifugal partition chromatography (CPC). Hydrodynamic instruments are often marketed as high-speed or high-performance countercurrent chromatography (HSCCC and HPCCC respectively) instruments which rely on the Archimedes' screw force in a helical coil to retain the stationary phase in the column.

The components of a CCC system are similar to most liquid chromatography configurations, such as high-performance liquid chromatography. One or more pumps deliver the phases to the column which is the CCC instrument itself. Samples are introduced into the column through a sample loop filled with an automated or manual syringe. The outflow is monitored with various detectors such as ultraviolet–visible spectroscopy or mass spectrometry. The operation of the pumps, CCC instrument, sample injection, and detection may be controlled manually or with a microprocessor.

Support-free Liquid Chromatography

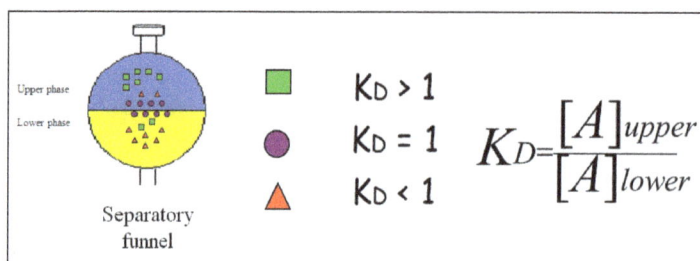

Partition coefficient (K_D).

Standard column chromatography consists of a solid stationary phase and a liquid mobile phase, while gas chromatography (GC) uses a solid or liquid stationary phase on a solid support and a gaseous mobile phase. By contrast, in liquid-liquid chromatography, both the mobile and stationary phases are liquid. The contrast is, however not as stark as it first appears. In reversed-phase chromatography, for example, the stationary phase can be regarded as a liquid which is immobilized by chemical bonding to a micro-porous silica solid support. In countercurrent chromatography centrifugal or gravitational forces immobilize the stationary liquid layer. By eliminating solid supports, permanent adsorption of the analyte onto the column is avoided, and a high recovery of the analyte can be achieved. The countercurrent chromatography instrument is easily switched between normal phase chromatography and reversed-phase chromatography simply by changing the mobile and stationary phases. With column chromatography, the separation potential is limited by the commercially available stationary phase media and its particular characteristics. Nearly any pair of immiscible solutions can be used in countercurrent chromatography provided that the stationary phase can be successfully retained.

Solvent costs are also generally lower than for high-performance liquid chromatography (HPLC). In comparison to column chromatography, flows and total solvent usage can in most countercurrent chromatography separations may be reduced by half and even up to a tenth. Also, the cost of purchasing and disposing of stationary phase media is eliminated. Another advantage of countercurrent chromatography is that experiments conducted in the laboratory can be scaled to industrial volumes. When gas chromatography or HPLC is carried out with large volumes, resolution is lost due to issues with surface-to-volume ratios and flow dynamics; this is avoided when both phases are liquid.

The CCC separation process can be thought of as occurring in three stages: mixing, settling, and separation of the two phases (although they often occur continuously). Vigorous mixing of the phases is critical in order to maximize the interfacial area between them and enhance mass transfer. The analyte will distribute between the phases according to its partition coefficient which is also called the distribution coefficient, distribution constant, or partition ratio and is represented by P, K, D, K_c, or K_D. The partition coefficient for an analyte in a particular biphasic solvent system is independent of the volume of the instrument, flow rate, stationary phase retention volume ratio and the g-force required to immobilize the stationary phase. The degree of stationary phase retention is a crucial parameter. Common factors that influence stationary phase retention are flow rate, solvent composition of the biphasic solvent system, and the g-force. The stationary phase retention is represented by the stationary phase volume retention ratio (Sf) which is the volume of the stationary phase divided by the total volume of the instrument. The settling time is a property of the solvent system and the sample matrix, both of which greatly influence stationary phase retention.

To most process chemists, the term "countercurrent" implies two immiscible liquids moving in opposing directions, as typically occurs in large centrifugal extractor units. With the exception of dual flow CCC, most countercurrent chromatography modes of operation have a stationary phase and a mobile phase. Even in this situation, countercurrent flows occur within the instrument column. Several researchers have proposed renaming both CCC & CPC to liquid-liquid chromatography, but others feel the term "countercurrent" itself is a misnomer.

Unlike column chromatography and high-performance liquid chromatography, countercurrent chromatography operators can inject large volumes relative to column volume. Typically 5 to 10% of coil volume can be injected. In some cases this can be increased to as high as 15 to 20% of the coil volume. Typically, most modern commercial CCC and CPC can inject 5 to 40 g per liter capacity. The range is so large, even for a specific instrument, let alone all instrument options, as the type of target, matrix and available biphasic solvent vary so much. Approximately 10 g per liter would be a more typical value, that the majority of applications could use as a base value.

The countercurrent separation starts with choosing an appropriate biphasic solvent system for the desired separation. A wide array of biphasic solvent mixtures are available to the CCC practitioner including the combination n-hexane (or heptane), ethyl acetate, methanol and water in different proportions. This basic solvent system is sometimes referred to as the HEMWat solvent system. The choice of solvent system may be guided by perusal of the CCC literature. The familiar technique of thin layer chromatography may also be employed to determine an optimal solvent system. The organization of solvent systems into "families" has greatly facilitated the choice of solvent systems as well. A solvent system can be tested with a one-flask partitioning experiment. The measured partition coefficient from the partitioning experiment will indicate the elution behavior of the compound. Typically, it is desirable to choose a solvent system where the target compound(s) have a partition coefficient between 0.25 and 8. Historically, it was thought that no commercial countercurrent chromatograph could cope with the high viscosities of ionic liquids. However, modern instruments that can accommodate 30 to 70+ % ionic liquids (and potentially 100% ionic liquid, if both phases are suitably customized ionic liquids) have become available. Ionic liquids can be customized for polar/non-polar organic, achiral and chiral compounds, bio-molecule, and inorganic separations, as ionic liquids can be customized to have extraordinary solvency and specificity.

After the biphasic solvent system has been chosen a batch of is formulated and equilibrated in a separatory funnel. This step is called pre-equilibration of the solvent system. The two phases are separated. Then the column is filled with stationary with a pump. Next, the column is set an equilibration conditions, such as the desired rotation speed, and the mobile phase is pumped through the column. The mobile phase displaces the a portion of the stationary phase until column equilibration is achieved and the mobile phase elutes from the column. The sample may be introduced into the column at any time during the column equilibration step or after equilibration has been accomplished. After the volume of eluant surpasses the volume of the mobile phase in the column, the sample components will begin to elute. Compounds with a partition coefficient of unity will elute when one column volume of mobile phase has passed through the column since the time of injection. The compound can then be introduced to another stationary phase to help increase the resolution of results. The flow is stopped after the target compound(s) are eluted or the column is extruded by pumping the stationary phase through the column. An example of a major application of countercurrent chromatography is to take an extremely complex matrix such as a plant extract, perform the countercurrent chromatography separation with a carefully selected solvent system, and extrude the column to recover all of the sample. The original complex matrix will have been fractionated into discrete narrow polarity bands, which may then be assayed for chemical composition or bioactivity. Performing one or more countercurrent chromatography separations in conjunction with other chromatographic and non chromatographic techniques has the potential for rapid advances in compositional recognition of extremely complex matrices.

Droplet CCC

Droplet countercurrent chromatography (DCCC) was introduced in 1970 by Tanimura, Pisano, Ito, and Bowman. DCCC uses only gravity to move the mobile phase through the stationary phase which is held in long vertical tubes connected in series. In the descending mode, droplets of the denser mobile phase and sample are allowed to fall through the columns of the lighter stationary phase using only gravity. If a less-dense mobile phase is used it will rise through the stationary phase; this is called ascending mode. The eluent from one column is transferred to another; the more columns that are used, the more theoretical plates can be achieved. DCCC enjoyed some success with natural product separations but was largely eclipsed by the rapid development of high-speed countercurrent chromatography. The main limitation of DCCC is that flow rates are low, and poor mixing is achieved for most binary solvent systems.

Hydrodynamic CCC

The modern era of CCC began with the development of the planetary centrifuge by Dr. Yoichiro Ito which was first introduced in 1966 as a closed helical tube which was rotated on a "planetary" axis as is turned on a "sun" axis. A flow-through model was subsequently developed and the new technique was called countercurrent chromatography in 1970. The technique was further developed by employing test mixtures of DNP amino acids in a chloroform:glacial acetic acid:0.1 M aqueous hydrochloric acid (2:2:1 v/v) solvent system. Much development was needed to engineer the instrument so that required planetary motion could be sustained while the phases were being pumped through the coil(s). Parameters such as the relative rotation of the two axes (synchronous or non-synchronous), the direction of flow through the coil, and the rotor angles were investigated.

High-speed

By 1982 the technology was sufficiently advanced for the technique to be called "high-speed" countercurrent chromatography (HSCCC). Peter Carmeci initially commercialized the PC Inc. Ito Multilayer Coil Separator/Extractor which utilized a single bobbin (onto which the coil is wound) and a counterbalance, plus a set of "flying leads" which are tubing that connect the bobbins. Dr. Walter Conway & others later evolved the bobbin design such that multiple coils, even coils of different tubing sizes, could be placed on the single bobbin. Edward Chou later evolved and commercialized a triple bobbin design as the Pharmatech CCC which had a de-twist mechanism for leads between the three bobbins. The Quattro CCC released in 1993 further evolved the commercially available instruments by utilizing a novel mirror image, twin bobbin design that did not need the de-twist mechanism of the Pharmatech between the multiple bobbins, so could still accommodate multiple bobbins on the same instrument. Hydrodynamic CCC are now available with up to 4 coils per instrument. These coils can be in PTFE, PEEK, PVDF, or stainless steel tubing. The 2, 3 or 4 coils can all be of the same bore to facilitate "2D" CCC. The coils may be connected in series to lengthen the coil and increase the capacity, or the coils may be linked in parallel so that 2, 3, or 4 separations may be done simultaneously. The coils can also be of different sizes, on one instrument, ranging from 1 to 6 mm on one instrument, thus allowing a single instrument to optimize from mg to kilos per day. More recently instrument derivatives have been offered with rotating seals for various hydrodynamic CCC designs, instead of flying leads, either as custom or standard options.

High-performance

The operating principle of CCC equipment requires a column consisting of a tube coiled around a bobbin. The bobbin is rotated in a double-axis gyratory motion (a cardioid), which causes a variable g-force to act on the column during each rotation. This motion causes the column to see one partitioning step per revolution and components of the sample separate in the column due to their partitioning coefficient between the two immiscible liquid phases. "High-performance" countercurrent chromatography (HPCCC) works in much the same way as HSCCC. A seven-year R&D process produced HPCCC instruments that generated 240 g's, compared to the 80 g's of the HSCCC machines. This increase in g-force and larger bore of the column has enabled a ten-fold increase in throughput, due to improved mobile phase flow rates and a higher stationary phase retention. Countercurrent chromatography is a preparative liquid chromatography technique, however with the advent of the higher-g HPCCC instruments it is now possible to operate instruments with sample loadings as low as a few milligrams, whereas in the past 100s of milligrams had been necessary. Major application areas for this technique include natural product purification and drug development.

Hydrostatic CCC

Hydrostatic CCC or centrifugal partition chromatography (CPC) was invented in the 1980s by the Japanese company Sanki Engineering Ltd, whose president was Kanichi Nunogaki. CPC has been extensively developed in France starting from the late 1990s. In France, they initially optimized the stacked disc concept initiated by Sanki. More recently, in France and UK, non-stacked disc CPC configurations have been developed with PTFE, stainless steel or titanium rotors. These have been designed to overcome possible leakages between the stacked discs of the original concept, and to allow steam cleaning for Good manufacturing practice. The volumes ranging from a 100 ml to 12 liters are available in different rotor materials. The 25-liter rotor CPC has a titanium rotor. This technique is sometimes sold under the name "fast" CPC or "high-performance" CPC.

Realization

The centrifugal partition chromatograph instrument is constituted with a unique rotor which contains the column. This rotor rotates on its central axis (while HSCCC column rotates on its planetary axis and simultaneously rotates eccentrically about another solar axis). With less vibrations and noise, the CPC offers a typical rotation speed range from 500 to 2000 rpm. Contrary to hydrodynamic CCC, the rotation speed is not directly proportional to the retention volume ratio of the stationary phase. Like DCCC, CPC can be operated in either descending or ascending mode, where the direction is relative to the force generated by the rotor rather than gravity. A redesigned CPC column with larger chambers and channels has been named centrifugal partition extraction (CPE). In the CPE design, faster flow rates and increased column loading can be achieved.

Advantages

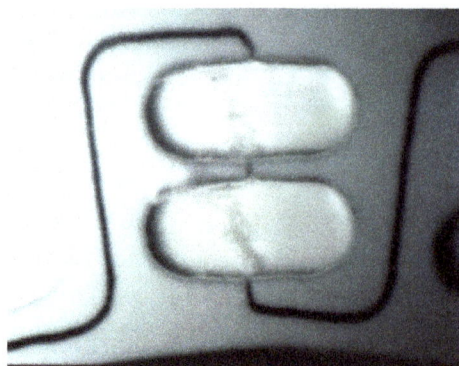

Visualization of mixing and settling in a CPC twin cell.

CPC offers direct scale-up from analytical apparatuses (few milliliters) to industrial apparatuses (several liters) for fast batch-production. CPC seems particularly suited to accommodate aqueous two-phase solvent systems. Generally, CPC instruments can retain solvent systems that are not well-retained in a hydrodynamic instrument due to small differences in density between the phases. It has been very helpful for the development of CPC instrumentation to visualize the flow patterns which give rise to the mixing and settling in the CPC chamber with an asynchronous camera and a stroboscope triggered by the CPC rotor.

Modes of Operation

The hydrodynamic and hydrostatic instruments may be employed in a variety of ways, or modes of operation, in order to address the particular separation needs of the scientist. Many modes of operation have been devised to take advantage of the strengths and potentialities of the countercurrent chromatography technique. Generally, the following modes may be performed with commercially available instruments.

Normal-phase

Normal phase elution is achieved by pumping the non-aqueous or phase of a biphasic solvent system through the column as the mobile phase, with a more polar stationary phase being retained in the column. The cause of original nomenclature of is relevant. As original stationary phases of paper chromatography were superseded by more efficient materials such as diatomaceous earths (natural micro-porous silica) and followed by modern silica gel, the thin-layer chromatography stationary phase was polar (hydroxy groups attached to silica) and maximum retention was achieved with non-polar solvents such as n-hexane. Progressively more polar eluents were then used to move polar compounds up the plate. Various alkane bonded phases were tried with C18 becoming the most popular. Alkane chains were chemically bonded to the silica, and a reversal of the elution trend occurred. Thus a polar stationary became "normal" phase chromatography, and the non-polar stationary phase chromatography became "reversed" phase chromatography.

Reversed-phase

In reversed-phase (e.g. aqueous mobile phase) elution, the aqueous phase is used as the mobile phase with a less polar stationary phase. In countercurrent chromatography the same solvent system may be used in either normal or reversed phase mode simply by switching the direction of mobile phase flow through the column.

Elution-extrusion

The extrusion of stationary phase from the column at the end of a separation experiment by stopping rotation and pumping solvent or gas through the column was used by CCC practitioners before the term EECCC was suggested. In elution-extrusion mode (EECCC), The mobile phase is extruded after a certain point by switching the phase being pumped into the system whilst maintaining rotation. For example, if the separation has been initiated with the aqueous phase as the mobile phase at a certain point the organic phase is pumped through the column which effectively pushes out both phases that are present in the column at the time of switching. The complete sample is eluted in the order of polarity (either normal or reversed) without loss of resolution by diffusion. It requires only one column volume of solvent phase and leaves the column full of fresh stationary phase for the subsequent separation.

Gradient

The use of a solvent gradient is very well developed in column chromatography but is less common in CCC. A solvent gradient is produced by increasing (or decreasing) the polarity of the mobile phase during the separation to achieve optimal resolution across a wider range of polarities. For example, a methanol-water mobile phase gradient may be employed using heptane as the stationary phase. This is not possible with all biphasic solvent systems, due to excessive loss of stationary phase created by disruption the equilibrium conditions within the column. Gradients may either be produced in steps, or continuously.

Dual-mode

In dual-mode, the mobile and stationary phases are reversed part way through the separation experiment. This requires changing the phase being pumped through the column as well as the direction of flow. Dual-mode operation is likely to elute the entire sample from the column but the order of elution is disrupted by switching the phase and direction of flow.

Dual-flow

Dual-flow, also known as dual, countercurrent chromatography occurs when both phases are flowing in opposite directions inside the column. Instruments are available for dual-flow operation for both Hydrodynamic and hydrostatic CCC. Dual-flow countercurrent chromatography was first described by Yoichiro Ito in 1985 for foam CCC where gas-liquid separations were performed. Liquid–liquid separations soon followed. The countercurrent chromatography instrument must be modified so that both ends of the column have both inlet and outlet capabilities. This mode may accommodate continuous or sequential separations with the sample being introduced in the middle of the column or between two bobbins in a hydrodynamic instrument. A technique called intermittent countercurrent extraction (ICcE) is a quasi-continuous method where the flow of the phases is alternated "intermittently" between normal and reversed-phase elution so that the stationary phase also alternates.

Recycling and Sequential

Recycling chromatography is mode practiced in both HPLC and CCC. In recycling chromatography, the target compounds are reintroduced into the column after they elute. Each pass through the column increases the number of theoretical plates the compounds experience and enhances chromatographic resolution. Direct recycling must be done with an isocratic solvent system. With this mode, the eluant can be selectively re-chromatographed on the same or a different column in order to facilitate the separation. This process of selective recycling has been termed a "heart-cut" and is especially effective in purifying selected target compounds with some sacrificial loss of recovery. The process of re-separating selected fractions from one chromatography experiment with another chromatographic method has long been practiced by scientists. Recycling and sequential chromatography is a streamlined version of this process. In CCC, the separation characteristics of the column may be modified simply by changing the composition of the biphasic solvent system.

Ion-exchange and pH-zone-refining

In an conventional CCC experiment the biphasic solvent system is pre-equilibrated before the instrument is filled with the stationary phase and equilibrated with the mobile phase. An ion-exchange mode has been created by modifying both of the phases after pre-equilibration. Generally, an ionic displacer (or eluter) is added to mobile phase and an ionic retainer is added to the stationary phase. For example, the aqueous mobile phase may contain NaI as a displacer and the organic stationary phase may be modified with the quaternary ammonium salt called Aliquat 336 as a retainer. The mode known a pH-zone-refining is a type of ion-exchange mode that utilizes acids and/or bases as solvent modifiers. Typically, the analytes are eluted in an order determined by their pKa values. For example, 6 oxindole alkaloids were isolated from a 4.5g sample of Gelsemium elegans stem extract with a biphasic solvent system composed of hexane–ethyl acetate–methanol–water (3:7:1:9, v/v) where 10 mM triethylamine (TEA) was added to the upper organic stationary phase as a retainer and 10 mM hydrochloric acid (HCl) to the aqueous mobile phase as an eluter. Ion-exchange modes such as pH-zone-refining have tremendous potential because high sample loads can be achieved without sacrificing separation power. It works best with ionizable compounds such as nitrogen containing alkaloids or carboxylic acid containing fatty acids.

Applications

Countercurrent chromatography and related liquid-liquid separation techniques have been used on both industrial and laboratory scale to purify a wide variety of chemical substances. Separation realizations include proteins, DNA, Cannabidiol (CBD) from Cannabis Sativa antibiotics, vitamins, natural products, pharmaceuticals, metal ions, pesticides, enantiomers, polyaromatic hydrocarbons from environmental samples, active enzymes, and carbon nanotubes. Countercurrent chromatography is known for its high dynamic range of scalability: milligram to kilogram quantities purified chemical components may be obtained with this technique. It also has the advantage of accommodating chemically complex samples with undissolved particulates.

Natural products extracts are commonly highly complex mixtures of active compounds and consequently their purification becomes a particularly challenging task. The development of a purification protocol to extract a single active component from the many hundreds that are often present in the mixture is something that can take months or even years to achieve, thus it is important for the natural product chemist to have, at their disposal, a broad range of diverse purification techniques. Counter-current chromatography (CCC) is one such separation technique utilising two immiscible phases, one as the stationary phase (retained in a spinning coil by centrifugal forces) and the second as the mobile phase. The method benefits from a number of advantages when compared with the more traditional liquid–solid separation methods, such as no irreversible adsorption, total recovery of the injected sample, minimal tailing of peaks, low risk of sample denaturation, the ability to accept particulates, and a low solvent consumption. The selection of an appropriate two-phase solvent system is critical to the running of CCC since this is both the mobile and the stationary phase of the system. However, this is also by far the most time consuming aspect of the technique and the one that most inhibits its general take-up. In recent years, numerous natural product purifications have been published using CCC from almost every country across the globe.

Periodic Counter-current Chromatography

Periodic counter-current chromatography (PCC) is a method for running affinity chromatography in a quasi-continuous manner. Today, the process is mainly employed for the purification of antibodies in the biopharmaceutical industry as well as in research and development. When purifying antibodies, Protein A is used as affinity matrix. However, periodic counter-current processes can be applied to any affinity type chromatography.

Basic Principle

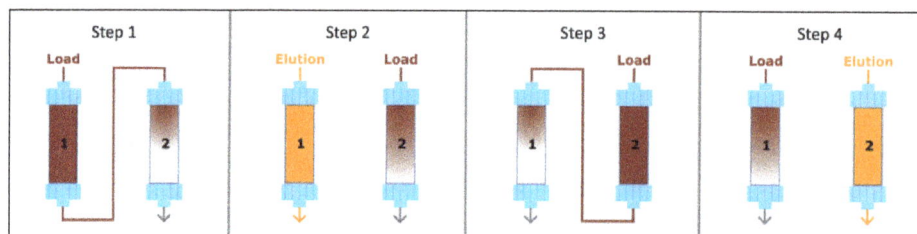

Process diagram for the 2-column periodic counter-current process "CaptureSMB".

In conventional affinity chromatography, a single chromatography column is loaded with feed material up to the point before target material (product) cannot be retained by the affinity material anymore. The resin with the adsorbed product on it is then washed to remove impurities. Finally, the pure product is eluted with a different buffer. Notably, if too much feed material is loaded onto the column, the product can break through and product is consequently lost. Therefore, it is very important to only partially load the column to maximize the yield.

Periodic counter-current chromatography puts this problem aside by utilizing more than one column. PCC processes can be run with any number of columns, starting from two. The following paragraph will explain a two-column version of PCC, but other protocols with more columns rely on the same principles. A diagram depicting the individual process steps is shown. In Step 1, the so-called sequential loading phase, columns 1 and 2 are interconnected. Column 1 is fully loaded with sample (red) while its breakthrough is captured on column 2. In Step 2, column 1 is washed, eluted, cleaned and re-equilibrated while loading separately continues on column 2. In Step 3, after regeneration of column 1, the columns are again inter-connected and column 2 is fully loaded while its breakthrough is captured on column 1. Finally, in Step 4 column 2 is washed, eluted, cleaned and re-equilibrated while loading continues independently on column 1. This cyclic process is repeated in a continuous way.

Several variations of periodic counter-current chromatography with more than two columns exist. In these cases, additional columns are either placed within the feed stream during loading, having the same effect as using longer columns. Alternatively, additional columns can be kept in an unoccupied stand-by mode during loading. This mode offers additional assurance that the main process is not influenced by washing and cleaning protocols, albeit in practice this is rarely required. On the other hand, the underutilized columns reduce the theoretical maximum productivity for such processes. Generally, the advantages and disadvantages of different multi-column protocols are the subject of debate. However, without a doubt, compared to single column batch processes, periodic counter-current processes provide significantly increased productivity.

Dynamic Process Control

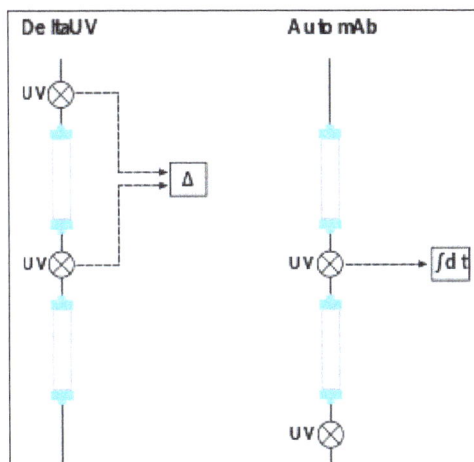

Dynamic process control mechanisms for periodic counter-current chromatography.

On the time scale of continuous chromatography runs, it is fairly common to observe changes in important process parameters, such as column health, buffer quality, feed titer (concentration) or feed composition. Such changes result in an altered maximum column capacity, relative to the amount of loaded feed material. In order to achieve a steady quality and yield for each process cycle, the timing of the individual process steps therefore has to be adjusted. Manual changes are in principle conceivable, but rather impractical. More commonly, dynamic process control algorithms monitor the process parameters and apply changes as needed automatically.

There are two different operating modes for dynamic process controllers in use today. The first one, called DeltaUV, monitors the difference between two signals from detectors situated before and after the first column. During initial loading, there is a large difference between the two signals, but it is diminishing as the impurities make their way through the column. Once the column is fully saturated with impurities and only additional product is being held back, the difference between the signals reaches a constant value. As long as the product is completely being captured on the column, the difference between the signals will remain constant. As soon as some of the product breaks through the column, the difference diminishes. Thus, the timing and amount of product breakthrough can be determined. The second possibility, called AutomAb, requires only the signal of a single detector situated behind the first column. During initial loading, the signal increases, as more and more impurities make their way through the column. When the column is saturated with impurities and as long as the product is completely being captured on the column, the signal then remains constant. As soon as some of the product breaks through the column, the signal increases again. Thus, the timing and amount of product breakthrough can again be determined.

Both iterations work equally well in theory. In practice, the requirement for two synced signals and the exposure of one detector to unpurified feed material, makes the DetaUV approach less reliable than AutomAb.

Commercial Situation

As of 2017, GE Healthcare holds patents around three-column periodic counter-current chromatography: this technology is used in their Äkta PCC instrument. Likewise, ChromaCon holds patents for an optimized two-column version (CaptureSMB). CaptureSMB is used in ChromaCon's Contichrom CUBE and under license in YMC's Ecoprime Twin systems. Additional manufacturers of systems capable of periodic counter-current chromatography include Novasep and Pall.

Ion Chromatography

Ion chromatography (or ion-exchange chromatography) is a chromatography process that separates ions and polar molecules based on their affinity to the ion exchanger. It works on almost any kind of charged molecule—including large proteins, small nucleotides, and amino acids. However, ion chromatography must be done in conditions that are one unit away from the isoelectric point of a protein.

The two types of ion chromatography are anion-exchange and cation-exchange. Cation-exchange chromatography is used when the molecule of interest is positively charged. The molecule is positively charged because the pH for chromatography is less than the pI. In this type of chromatography, the stationary phase is negatively charged and positively charged molecules are loaded to be attracted to it. Anion-exchange chromatography is when the stationary phase is positively charged and negatively charged molecules (meaning that pH for chromatography is greater than the pI) are loaded to be attracted to it. It is often used in protein purification, water analysis, and quality control. The water-soluble and charged molecules such as proteins, amino acids, and peptides bind to moieties which are oppositely charged by forming ionic bonds to the insoluble stationary phase. The equilibrated stationary phase consists of an ionizable functional group where the targeted molecules of a mixture to be separated and quantified can bind while passing through the column—a cationic stationary phase is used to separate anions and an anionic stationary phase is used to separate cations. Cation exchange chromatography is used when the desired molecules to separate are cations and anion exchange chromatography is used to separate anions. The bound molecules then can be eluted and collected using an eluant which contains anions and cations by running higher concentration of ions through the column or changing pH of the column.

One of the primary advantages for the use of ion chromatography is only one interaction involved during the separation as opposed to other separation techniques; therefore, ion chromatography may have higher matrix tolerance. Another advantage of ion exchange, is the predictability of elution patterns (based on the presence of the ionizable group). For example, when cation exchange chromatography is used, cations will elute out last. Meanwhile, the negative charged molecules will elute out first. However, there are also disadvantages involved when performing ion-exchange chromatography, such as constant evolution with the technique which leads to the inconsistency from column to column. A major limitation to this purification technique is that it is limited to ionizable group.

Principle

Ion-exchange chromatography separates molecules based on their respective charged groups. Ion-exchange chromatography retains analyte molecules on the column based on coulombic (ionic) interactions. The ion exchange chromatography matrix consists of positively and negatively charged ions.

Essentially, molecules undergo electrostatic interactions with opposite charges on the stationary phase matrix. The stationary phase consists of an immobile matrix that contains charged ionizable functional groups or ligands. The stationary phase surface displays ionic functional groups (R-X) that interact with analyte ions of opposite charge. To achieve electroneutrality, these inert charges couple with exchangeable counterions in the solution. Ionizable molecules that are to be purified compete with these exchangeable counterions for binding to the immobilized charges on the stationary phase. These ionizable molecules are retained or eluted based on their charge. Initially, molecules that do not bind or bind weakly to the stationary phase are first to wash away. Altered conditions are needed for the elution of the molecules that bind to the stationary phase. The concentration of the exchangeable counterions, which competes with the molecules for binding, can be increased or the pH can be changed. A change in pH affects the charge on the particular molecules and, therefore, alters binding. The molecules then start eluting out based on the changes in their charges from the adjustments. Further such adjustments can be used to release the protein of interest. Additionally, concentration of counterions can be gradually varied to separate ionized molecules. This type of elution is called gradient elution. On the other hand, step elution can be used in which the concentration of counterions are varied in one step. This type of chromatography is further subdivided into cation exchange chromatography and anion-exchange chromatography. Positively charged molecules bind to cation exchange resins while negatively charged molecules bind to anion exchange resins. The ionic compound consisting of the cationic species M^+ and the anionic species B^- can be retained by the stationary phase.

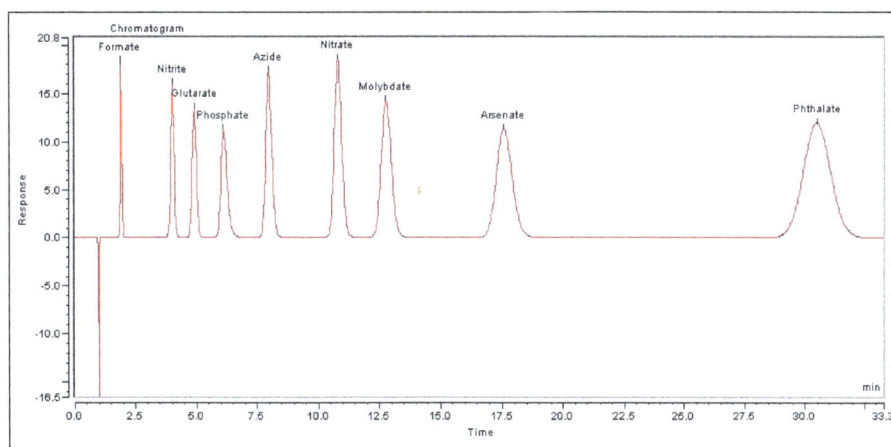

Ion chromatogram displaying anion separation.

Cation exchange chromatography retains positively charged cations because the stationary phase displays a negatively charged functional group:

$$R\text{-}X^-C^+ + M^+B^- \rightleftharpoons R\text{-}X^-M^+ + C^+ + B^-$$

Anion exchange chromatography retains anions using positively charged functional group:

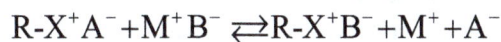

$$R\text{-}X^+A^- + M^+B^- \rightleftharpoons R\text{-}X^+B^- + M^+ + A^-$$

The ion strength of either C^+ or A^- in the mobile phase can be adjusted to shift the equilibrium position, thus retention time.

The ion chromatogram shows a typical chromatogram obtained with an anion exchange column.

Procedure

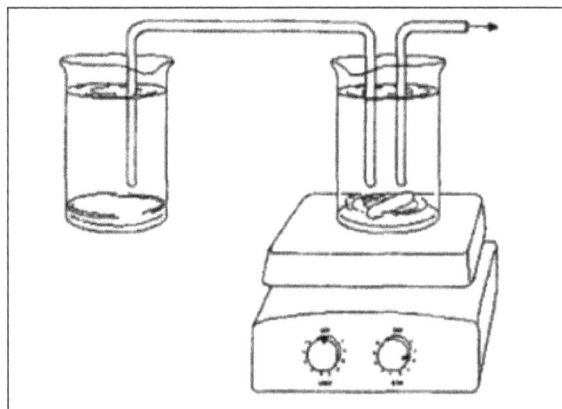

Chamber (left) contains high salt concentration. Stirred chamber (right) contains low salt concentration. Gradual stirring causes the formation of a salt gradient as salt travel from high to low concentrations.

Before ion-exchange chromatography can be initiated, it must be equilibrated. The stationary phase must be equilibrated to certain requirements that depend on the experiment that you are working with. Once equilibrated, the charged ions in the stationary phase will be attached to its opposite charged exchangeable ions. Exchangeable ions such as Cl^- or Na^+. Next, a buffer should be chosen in which the desired protein can bind to. After equilibration, the column needs to be washed. The washing phase will help elute out all impurities that does not bind to the matrix while the protein of interest remains bounded. This sample buffer needs to have the same pH as the buffer used for equilibration to help bind the desired proteins. Uncharged proteins will be eluted out of the column at a similar speed of the buffer flowing through the column. Once the sample has been loaded onto to the column and the column has been washed with the buffer to elute out all non-desired proteins, elution is carried out to elute the desired proteins that are bound to the matrix. Bound proteins are eluted out by utilizing a gradient of linearly increasing salt concentration. With increasing ionic strength of the buffer, the salt ions will compete with the desired proteins in order to bind to charged groups on the surface of the medium. This will cause desired proteins to be eluted out of the column.

Proteins that have a low net charge will be eluted out first as the salt concentration increases causing the ionic strength to increase. Proteins with high net charge will need a higher ionic strength for them to be eluted out of the column. It is possible to perform ion exchange chromatography in bulk, on thin layers of medium such as glass or plastic plates coated with a layer of the desired stationary phase, or in chromatography columns. Thin layer chromatography or column chromatography share similarities in that they both act within the same governing principles; there is constant and frequent exchange of molecules as the mobile phase travels along the stationary phase. It is not imperative to add the sample in minute volumes as the predetermined conditions for the exchange column have been chosen so that there will be strong interaction between the mobile and stationary phases. Furthermore, the mechanism of the elution process will cause a compartmentalization of the differing molecules based on their respective chemical characteristics. This phenomenon is due to an increase in salt concentrations at or near the top of the column, thereby displacing the molecules at that position, while molecules bound lower are released at a later point when the higher salt concentration reaches that area. These principles are the reasons that ion exchange chromatography is an excellent candidate for initial chromatography steps in a complex purification procedure as it can quickly yield small volumes of target molecules regardless of a greater starting volume.

Comparatively simple devices are often used to apply counterions of increasing gradient to a chromatography column. Counterions such as copper (II) are chosen most often for effectively separating peptides and amino acids through complex formation.

A simple device can be used to create a salt gradient. Elution buffer is consistently being drawn from the chamber into the mixing chamber, thereby altering its buffer concentration. Generally, the buffer placed into the chamber is usually of high initial concentration, whereas the buffer placed into the stirred chamber is usually of low concentration. As the high concentration buffer from the left chamber is mixed and drawn into the column, the buffer concentration of the stirred column gradually increase. Altering the shapes of the stirred chamber, as well as of the limit buffer, allows for the production of concave, linear, or convex gradients of counterion.

A multitude of different mediums are used for the stationary phase. Among the most common immobilized charged groups used are trimethylaminoethyl (TAM), triethylaminoethyl (TEAE), diethyl-2-hydroxypropylaminoethyl (QAE), aminoethyl (AE), diethylaminoethyl (DEAE), sulpho (S), sulphomethyl (SM), sulphopropyl (SP), carboxy (C), and carboxymethyl (CM).

Successful packing of the column is an important aspect of ion chromatography. Stability and efficiency of a final column depends on packing methods, solvent used, and factors that affect mechanical properties of the column. In contrast to early inefficient dry-packing methods, wet slurry packing, in which particles that are suspended in an appropriate solvent are delivered into a column under pressure, shows significant improvement. Three different approaches can be employed in performing wet slurry packing: the balanced density method (solvent's density is about that of porous silica particles), the high viscosity method (a solvent of high viscosity is used), and the low viscosity slurry method (performed with low viscosity solvents).

Polystyrene is used as a medium for ion-exchange. It is made from the polymerization of styrene with the use of divinylbenzene and benzoyl peroxide. Such exchangers form hydrophobic interactions with proteins which can be irreversible. Due to this property, polystyrene ion exchangers are not suitable for protein separation. They are used on the other hand for the separation of small molecules in amino acid separation and removal of salt from water. Polystyrene ion exchangers with large pores can be used for the separation of protein but must be coated with a hydrophillic substance.

Cellulose based medium can be used for the separation of large molecules as they contain large pores. Protein binding in this medium is high and has low hydrophobic character. DEAE is an anion exchange matrix that is produced from a positive side group of diethylaminoethyl bound to cellulose or Sephadex.

Agarose gel based medium contain large pores as well but their substitution ability is lower in comparison to dextrans. The ability of the medium to swell in liquid is based on the cross-linking of these substances, the pH and the ion concentrations of the buffers used.

Incorporation of high temperature and pressure allows a significant increase in the efficiency of ion chromatography, along with a decrease in time. Temperature has an influence of selectivity due to its effects on retention properties. The retention factor ($k = (t_R{}^g - t_M{}^g)/(t_M{}^g - t_{ext})$) increases with temperature for small ions, and the opposite trend is observed for larger ions.

Despite ion selectivity in different mediums, further research is being done to perform ion exchange chromatography through the range of 40–175 °C.

An appropriate solvent can be chosen based on observations of how column particles behave in a solvent. Using an optical microscope, one can easily distinguish a desirable dispersed state of slurry from aggregated particles.

Weak and Strong Ion Exchangers

A "strong" ion exchanger will not lose the charge on its matrix once the column is equilibrated and so a wide range of pH buffers can be used. "Weak" ion exchangers have a range of pH values in which they will maintain their charge. If the pH of the buffer used for a weak ion exchange column goes out of the capacity range of the matrix, the column will lose its charge distribution and the molecule of interest may be lost. Despite the smaller pH range of weak ion exchangers, they are often used over strong ion exchangers due to their having greater specificity. In some experiments, the retention times of weak ion exchangers are just long enough to obtain desired data at a high specificity.

Resins (often termed 'beads') of ion exchange columns may include functional groups such as weak/strong acids and weak/strong bases. There are also special columns that have resins with amphoteric functional groups that can exchange both cations and anions. Some examples of functional groups of strong ion exchange resins are quaternary ammonium cation (Q), which is an anion exchanger, and sulfonic acid (S, $-SO_2OH$), which is a cation exchanger. These types of exchangers can maintain their charge density over a pH range of 0–14. Examples of functional groups of Weak ion exchange resins include diethylaminoethyl (DEAE, $-C_2H_4N(CH_2H_5)_2$), which is an anion exchanger, and carboxymethyl (CM, $-CH_2-COOH$), which is a cation exchanger. These two types of exchangers can maintain the charge density of their columns over a pH range of 5–9.

In ion chromatography, the interaction of the solute ions and the stationary phase based on their charges determines which ions will bind and to what degree. When the stationary phase features positive groups which attracts anions, it is called an anion exchanger; when there are negative groups on the stationary phase, cations are attracted and it is a cation exchanger. The attraction between ions and stationary phase also depends on the resin, organic particles used as ion exchangers.

Each resin features relative selectivity which varies based on the solute ions present who will compete to bind to the resin group on the stationary phase. The selectivity coefficient, the equivalent to the equilibrium constant, is determined via a ratio of the concentrations between the resin and each ion, however, the general trend is that ion exchangers prefer binding to the ion with a higher charge, smaller hydrated radius, and higher polarizability, or the ability for the electron cloud of an ion to be disrupted by other charges. Despite this selectivity, excess amounts of an ion with a lower selectivity introduced to the column would cause the lesser ion to bind more to the stationary phase as the selectivity coefficient allows fluctuations in the binding reaction that takes place during ion exchange chromatography.

Typical Technique

A sample is introduced, either manually or with an autosampler, into a sample loop of known volume. A buffered aqueous solution known as the mobile phase carries the sample from the loop onto a column that contains some form of stationary phase material. This is typically a resin or gel matrix consisting of agarose or cellulose beads with covalently bonded charged functional groups. Equilibration of the stationary phase is needed in order to obtain the desired charge of the column. If the column is not properly equilibrated the desired molecule may not bind strongly to the column. The target analytes (anions or cations) are retained on the stationary phase but can be eluted by increasing the concentration of a similarly charged species that displaces the analyte ions from the stationary phase. For example, in cation exchange chromatography, the positively charged analyte can be displaced by adding positively charged sodium ions. The analytes of interest must then be detected by some means, typically by conductivity or UV/visible light absorbance.

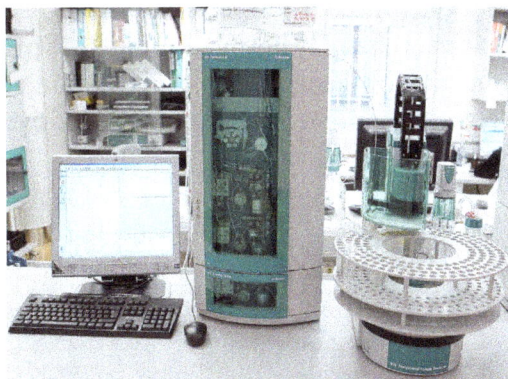
Metrohm 850 Ion chromatography system.

Control an IC system usually requires a chromatography data system (CDS). In addition to IC systems, some of these CDSs can also control gas chromatography (GC) and HPLC.

Membrane Exchange Chromatography

A type of ion exchange chromatography, membrane exchange is a relatively new method of purification designed to overcome limitations of using columns packed with beads. Membrane Chromatographic devices are cheap to mass-produce and disposable unlike other chromatography devices that require maintenance and time to revalidate. There are three types of membrane absorbers that are typically used when separating substances. The three types are flat sheet, hollow fibre, and radial flow. The most common absorber and best suited for membrane chromatography is multiple flat sheets because it has more absorbent volume. It can be used to overcome mass transfer limitations and pressure drop, making it especially advantageous for isolating and purifying viruses, plasmid DNA, and other large macromolecules. The column is packed with microporous membranes with internal pores which contain adsorptive moieties that can bind the target protein. Adsorptive membranes are available in a variety of geometries and chemistry which allows them to be used for purification and also fractionation, concentration, and clarification in an efficiency that is 10 fold that of using beads. Membranes can be prepared through isolation of the membrane itself, where membranes are cut into squares and immobilized. A more recent method involved the use of live cells that are attached to a support membrane and are used for identification and clarification of signaling molecules.

Separating Proteins

Ion exchange chromatography can be used to separate proteins because they contain charged functional groups. The ions of interest (in this case charged proteins) are exchanged for another ions (usually H⁺) on a charged solid support. The solutes are most commonly in a liquid phase, which tends to be water. Take for example proteins in water, which would be a liquid phase that is passed through a column. The column is commonly known as the solid phase since it is filled with porous synthetic particles that are of a particular charge. These porous particles are also referred to as beads, may be aminated (containing amino groups) or have metal ions in order to have a charge. The column can be prepared using porous polymers, for macromolecules over 100,000 the optimum size of the porous particle is about 1 μm2. This is because slow diffusion of the solutes within the pores does not restrict the separation quality. The beads containing positively charged groups, which attract the negatively charged proteins, are commonly referred to as anion exchange resins. The amino acids that have negatively charged side chains at pH 7 (pH of water) are glutamate and aspartate. The beads that are negatively charged are called cation exchange resins, as positively charged proteins will be attracted. The amino acids that have positively charged side chains at pH 7 are lysine, histidine and arginine.

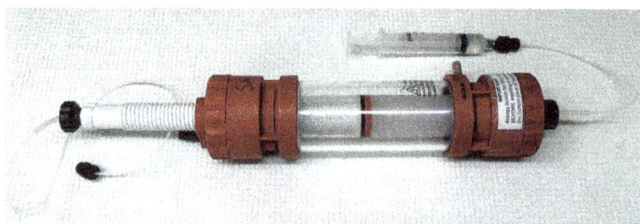

Preparative-scale ion exchange column used for protein purification.

The isoelectric point is the pH at which a compound - in this case a protein - has no net charge. A protein's isoelectric point or PI can be determined using the pKa of the side chains, if the amino (positive chain) is able to cancel out the carboxyl (negative) chain, the protein would be at its PI. Using buffers instead of water for proteins that do not have a charge at pH 7, is a good idea as it enables the manipulation of pH to alter ionic interactions between the proteins and the beads. Weakly acidic or basic side chains are able to have a charge if the pH is high or low enough respectively. Separation can be achieved based on the natural isoelectric point of the protein. Alternatively a peptide tag can be genetically added to the protein to give the protein an isoelectric point away from most natural proteins (e.g., 6 arginines for binding to a cation-exchange resin or 6 glutamates for binding to an anion-exchange resin such as DEAE-Sepharose).

Elution by increasing ionic strength of the mobile phase is more subtle. It works because ions from the mobile phase interact with the immobilized ions on the stationary phase, thus "shielding" the stationary phase from the protein, and letting the protein elute.

Elution from ion-exchange columns can be sensitive to changes of a single charge- chromatofocusing. Ion-exchange chromatography is also useful in the isolation of specific multimeric protein assemblies, allowing purification of specific complexes according to both the number and the position of charged peptide tags.

Gibbs–Donnan Effect

In ion exchange chromatography, the Gibbs–Donnan effect is observed when the pH of the applied buffer and the ion exchanger differ, even up to one pH unit. For example, in anion-exchange columns, the ion exchangers repeal protons so the pH of the buffer near the column differs is higher than the rest of the solvent. As a result, an experimenter has to be careful that the protein(s) of interest is stable and properly charged in the "actual" pH.

This effect comes as a result of two similarly charged particles, one from the resin and one from the solution, failing to distribute properly between the two sides; there is a selective uptake of one ion over another. For example, in a sulphonated polystyrene resin, a cation exchange resin, the chlorine ion of a hydrochloric acid buffer should equilibrate into the resin. However, since the concentration of the sulphonic acid in the resin is high, the hydrogen of HCl has no tendency to enter the column. This, combined with the need of electroneutrality, leads to a minimum amount of hydrogen and chlorine entering the resin.

Uses

Clinical Utility

A use of ion chromatography can be seen in the argentation ion chromatography. Usually, silver and compounds containing acetylenic and ethylenic bonds have very weak interactions. This phenomenon has been widely tested on olefin compounds. The ion complexes the olefins make with silver ions are weak and made based on the overlapping of pi, sigma, and d orbitals and available electrons therefore cause no real changes in the double bond. This behavior was manipulated to separate lipids, mainly fatty acids from mixtures in to fractions with differing number of double bonds using silver ions. The ion resins were impregnated with silver ions, which were then exposed to various acids (silicic acid) to elute fatty acids of different characteristics.

Detection limits as low as 1 μM can be obtained for alkali metal ions. It may be used for measurement of HbA1c, porphyrin and with water purification. Ion Exchange Resins(IER) have been widely used especially in medicines due to its high capacity and the uncomplicated system of the separation process. One of the synthetic uses is to use Ion Exchange Resins for kidney dialysis. This method is used to separate the blood elements by using the cellulose membraned artificial kidney.

Another clinical application of ion chromatography is in the rapid anion exchange chromatography technique used to separate creatine kinase (CK) isoenzymes from human serum and tissue sourced in autopsy material (mostly CK rich tissues were used such as cardiac muscle and brain). These isoenzymes include MM, MB, and BB, which all carry out the same function given different amino acid sequences. The functions of these isoenzymes are to convert creatine, using ATP, into phosphocreatine expelling ADP. Mini columns were filled with DEAE-Sephadex A-50 and further eluted with tris-buffer sodium chloride at various concentrations (each concentration was chosen advantageously to manipulate elution). Human tissue extract was inserted in columns for separation. All fractions were analyzed to see total CK activity and it was found that each source of CK isoenzymes had characteristic isoenzymes found within. Firstly, CK- MM was eluted, then CK-MB, followed by CK-BB. Therefore, the isoenzymes found in each sample could be used to identify the source, as they were tissue specific.

Using the information from results, correlation could be made about the diagnosis of patients and the kind of CK isoenzymes found in most abundant activity. From the finding, about 35 out of 71 patients studied suffered from heart attack (myocardial infarction) also contained an abundant amount of the CK-MM and CK-MB isoenzymes. Findings further show that many other diagnosis including renal failure, cerebrovascular disease, and pulmonary disease were only found to have the CK-MM isoenzyme and no other isoenzyme. The results from this study indicate correlations between various diseases and the CK isoenzymes found which confirms previous test results using various techniques. Studies about CK-MB found in heart attack victims have expanded since this study and application of ion chromatography.

Industrial Applications

Since 1975 ion chromatography has been widely used in many branches of industry. The main beneficial advantages are reliability, very good accuracy and precision, high selectivity, high speed, high separation efficiency, and low cost of consumables. The most significant development related to ion chromatography are new sample preparation methods; improving the speed and selectivity of analytes separation; lowering of limits of detection and limits of quantification; extending the scope of applications; development of new standard methods; miniaturization and extending the scope of the analysis of a new group of substances. Allows for quantitative testing of electrolyte and proprietary additives of electroplating baths. It is an advancement of qualitative hull cell testing or less accurate UV testing. Ions, catalysts, brighteners and accelerators can be measured. Ion exchange chromatography has gradually become a widely known, universal technique for the detection of both anionic and cationic species. Applications for such purposes have been developed, or are under development, for a variety of fields of interest, and in particular, the pharmaceutical industry. The usage of ion exchange chromatography in pharmaceuticals has increased in recent years, and in 2006, a chapter on ion exchange chromatography was officially added to the United States Pharmacopia-National Formulary (USP-NF). Furthermore, in 2009 release of the USP-NF, the United States Pharmacopia made several analyses of ion chromatography available using two techniques: conductivity detection, as well as pulse amperometric detection. Majority of these applications are primarily used for measuring and analyzing residual limits in pharmaceuticals, including detecting the limits of oxalate, iodide, sulfate, sulfamate, phosphate, as well as various electrolytes including potassium, and sodium. In total, the 2009 edition of the USP-NF officially released twenty eight methods of detection for the analysis of active compounds, or components of active compounds, using either conductivity detection or pulse amperometric detection.

Drug Development

There has been a growing interest in the application of IC in the analysis of pharmaceutical drugs. IC is used in different aspects of product development and quality control testing. For example, IC is used to improve stabilities and solubility properties of pharmaceutical active drugs molecules as well as used to detect systems that have higher tolerance for organic solvents. IC has been used for the determination of analytes as a part of a dissolution test. For instance, calcium dissolution tests have shown that other ions present in the medium can be well resolved among themselves and also from the calcium ion. Therefore, IC has been employed in drugs in the form of tablets and capsules in order to determine the amount of drug dissolve with time. IC is also widely used for detection and quantification of excipients or inactive ingredients used in pharmaceutical formulations. Detection of sugar and sugar alcohol in such formulations through IC has been done due to these polar groups getting resolved in ion column. IC methodology also established in analysis of impurities in drug substances and products. Impurities or any components that are not part of the drug chemical entity are evaluated and they give insights about the maximum and minimum amounts of drug that should be administered in a patient per day.

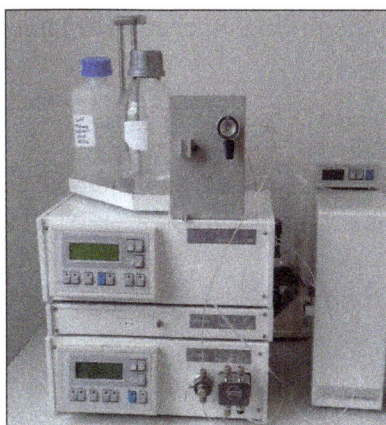

An ion chromatography system used to detect and measure cations such as sodium, ammonium and potassium in Expectorant Cough Formulations.

Ion chromatography (IC) is applied for separation and analysis of both anions and cations in environmental samples. The separation of analytes on a column leads to the identical analytical results as in other separation techniques. The differences in ion-exchange affinities of different species of ions result in their separation, and the most common detection is through conductivity detectors. The most widely used approach is the ion (cation) exchange with reversible complexation. However, for complex samples anion exchange with irreversible complexation is also employed. But nowadays the most effective one is the bifunctional ion exchange column that is mostly utilized to separate heavy and transition metals. Application of the chelation IC, and the online sample pretreatment quantification of lead, copper, cadmium, and other transition metals, have been carried out in water samples. The analysis of drinking water resulted in quantities of transition metals and heavy metals in the ng mL−1 range such as Pb (2.0), Cu (0.2), Cd (0.6), Co (0.2), and Ni (0.2).

Basic Process of IC

The basic process of chromatography using ion exchange can be represented in 5 steps: eluent loading, sample injection, separation of sample, elution of analyte A, and elution of analyte B. Elution is the process where the compound of interest is moved through the column. This happens because the eluent, the solution used as the solvent in chromatography, is constantly pumped through the column. The chemical reactions below are for an anion exchange process.

Step 1: The eluent loaded onto the column displaces any anions bonded to the resin and saturates the resin surface with the eluent anion.

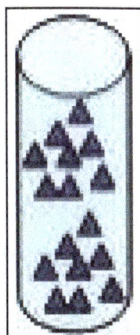

(Key: Eluent ion = ▲, Ion A= 🟨, Ion B = 🔴).

This process of the eluent ion (E^-) displacing an anion (X^-) bonded to the resin can be expressed by the following chemical reaction:

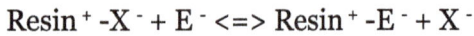

$$Resin^+ - X^- + E^- <=> Resin^+ - E^- + X^-$$

Step 2: A sample containing anion A and anion B are injected onto the column. This sample could contain many different ions, but for simplicity this example uses just two different ions ready to be injected onto the column.

Step 3: After the sample has been injected, the continued addition of eluent causes a flow through the column. As the sample elutes (or moves through the column), anion A and anion B adhere to the column surface differently. The sample zones move through the column as eluent gradually displaces the analytes.

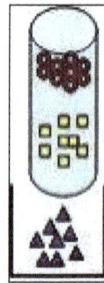

Step 4: The continued addition of the eluent causes a flow through the column. As sample elutes, anion A and anion B adhere to the column surface differently. The sample zones move through the column as eluent gradually displaces the analytes.

In reality not every eluent ion is removed from the surface of the column. It depends on the amount of analyte loaded. A better representation of the column can be seen by just looking at a slice of the column where the separation is occurring, as shown in the figure.

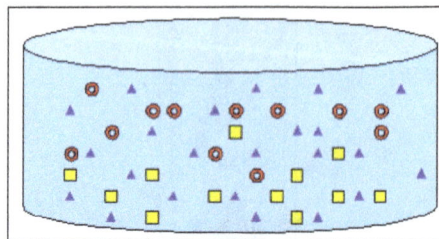

Step 5: As the eluent continues to be added, the anion A moves through the column in a band and ultimately is eluted first.

This process can be represented by the chemical reaction showing the displacement of the bound anion (A^-) by the eluent anion (E^-).

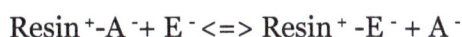

$$Resin^+\text{-}A^- + E^- <=> Resin^+\text{-}E^- + A^-$$

Step 6: The eluent displaces anion B, and anion B is eluted off the column.

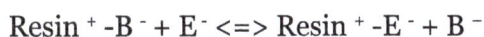

$$Resin^+\text{-}B^- + E^- <=> Resin^+\text{-}E^- + B^-$$

The overall 6 step process can be represented pictorally:

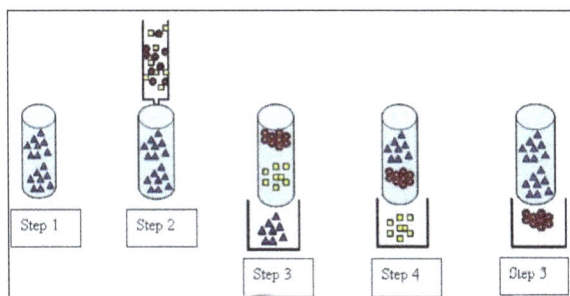

Stationary Phase (or Resin) Composition

There are a number of different resins or stationary phases that have been developed for use in IC. The main classes of substances used are: modified organic polymer resins, modified silica gels, inorganic salts, glasses, zeolites, metal oxides, and cellulose derivatives. The most commonly used resins are the silica gels and polymer resins. As the sample is injected onto the column, the two different analytes briefly displace the eluent as the counter-ion to the charged resin. The analyte is briefly retained at the fixed charge on the resin surface. The analytes are subsequently displaced by the eluent ions as the eluent is added to the column. The different affinities are the basis for the separation. The K_f values of each reaction is also known as the selectivity coefficient. The greater the difference between the K_f values for the two analytes, the more the two analytes will be separated during the ion chromatography process. In reality, the interaction between the solvent and the analyte can also have an impact on the order each analyte is eluted.

The common cation exchange resins are based on either polystyrenedivinylbenzene (PS-DVB) or methacrylate polymers. The surface of these polymers is functionalized with a negatively charged sulfonated group ($-SO_3^-$). The cation in the eluent or the analyte of interest is the counter-ion in the vicinity of the charged functional group.

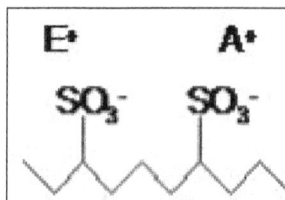

Cation exchange surface.

The surface of the polymer is functionalize with a quaternary amine ($-N^+R_3$) for anion exchange. The quaternary amine provides a positive charge to the surface, attracting negatively charged anions in the liquid phase. Just like the cation exchange resin, the anion of the eluent or the analyte of interest exists as the counter-ion in the vicinity of the positive charge residing on the amine.

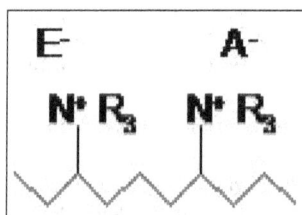

Anion exchange surface. The R stands for some organic (C and H) chain.

Electrochromatography

Electrochromatography is a chemical separation technique in analytical chemistry, biochemistry and molecular biology used to resolve and separate mostly large biomolecules such as proteins. It is a combination of size exclusion chromatography (gel filtration chromatography) and gel electrophoresis. These separation mechanisms operate essentially in superposition along the length of a gel filtration column to which an axial electric field gradient has been added. The molecules are separated by size due to the gel filtration mechanism and by electrophoretic mobility due to the gel electrophoresis mechanism. Additionally there are secondary chromatographic solute retention mechanisms.

Capillary Electrochromatography

Capillary electrochromatography (CEC) is a chromatographic technique in which the mobile phase is driven through the chromatographic bed by electroosmosis. Capillary electrochromatography is a combination of two analytical techniques, high-performance liquid chromatography and capillary electrophoresis. Capillary electrophoresis aims to separate analytes on the basis of their mass-to-charge ratio by passing a high voltage across ends of a capillary tube, which is filled with the analyte. High-performance liquid chromatography separates analytes by passing them, under high pressure, through a column filled with stationary phase. The interactions between the analytes and the stationary phase and mobile phase lead to the separation of the analytes. In capillary electrochromatography capillaries, packed with HPLC stationary phase, are subjected to a high voltage. Separation is achieved by electrophoretic migration of solutes and differential partitioning.

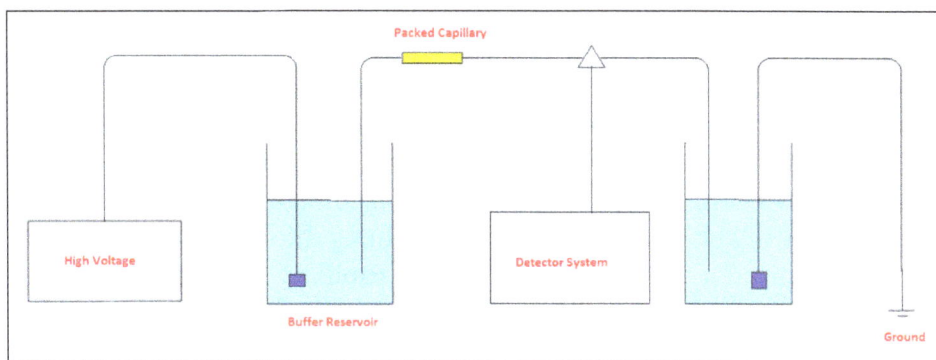

Mechanism of capillary electrochromatography.

Principle

Capillary electrochromatography (CEC) combines the principles used in HPLC and CE. The mobile phase is driven across the chromatographic bed using electroosmosis instead of pressure (as in HPLC). Electroosmosis is the motion of liquid induced by an applied potential across a porous material, capillary tube, membrane or any other fluid conduit. Electroosmotic flow is caused by the Coulomb force induced by an electric field on net mobile electric charge in a solution. Under alkaline conditions, the surface silanol groups of the fused silica will become ionised leading to a negatively charged surface. This surface will have a layer of positively charged ions in close proximity which are relatively immobilised. This layer of ions is called the Stern layer. The thickness of the double layer is given by the formula:

$$\delta = \sqrt{\frac{\epsilon_r \epsilon_0 RT}{2cF^2}}$$

where, ϵ_r is the relative permittivity of the medium, ϵ_0 is the permittivity of vacuum, R is the universal gas constant, T is the absolute temperature, c is the molar concentration, and F is the Faraday constant.

When an electric field is applied to the fluid (usually via electrodes placed at inlets and outlets), the net charge in the electrical double layer is induced to move by the resulting Coulomb force. The resulting flow is termed electroosmotic flow. In CEC positive ions of the electrolyte added along with the analyte accumulate in the electrical double layer of the particles of the column packing on application of an electric field they move towards the cathode and drag the liquid mobile phase with them.

The relationship between the linear velocity u of the liquid in the capillary and the applied electric field is given by the Smoluchowski equation as,

$$u = \epsilon_r \epsilon_0 \zeta E \eta$$

where ζ is the potential across the Stern layer (zeta potential), E is the electric field strength, and η is the viscosity of the solvent.

Separation of components in CEC is based on interactions between the stationary phase and differential electrophoretic migration of solutes.

Instrumentation

The components of a capillary electrochromatograph are a sample vial, source and destination vials, a packed capillary, electrodes, a high voltage power supply, a detector, and a data output and handling device. The source vial, destination vial and capillary are filled with an electrolyte such as an aqueous buffer solution. The capillary is packed with stationary phase. To introduce the sample, the capillary inlet is placed into a vial containing the sample and then returned to the source vial (sample is introduced into the capillary via capillary action, pressure, or siphoning). The migration of the analytes is then initiated by an electric field that is applied between the source and destination vials and is supplied to the electrodes by the high-voltage power supply. The analytes separate as they migrate due to their electrophoretic mobility, and are detected near the outlet end of the capillary. The output of the detector is sent to a data output and handling device such as an integrator or computer. The data is then displayed as an electropherogram, which reports detector response as a function of time. Separated chemical compounds appear as peaks with different migration times in an electropherogram.

Advantages

Avoiding the use of pressure to introduce the mobile phase into the column, results in a number of important advantages. Firstly, the pressure driven flow rate across a column depends directly on the square of the particle diameter and inversely on the length of the column. This restricts the length of the column and size of the particle, particle size is seldom less than 3 micrometer and the length of the column is restricted to 25 cm. Electrically driven flow rate is independent of length of column and size. A second advantage of using electroosmosis to pass the mobile phase into the column is the plug-like flow velocity profile of EOF, which reduces the solute dispersion in the column, increasing column efficiency.

Displacement Chromatography

Displacement chromatography is a chromatography technique in which a sample is placed onto the head of the column and is then displaced by a solute that is more strongly sorbed than the components of the original mixture. The result is that the components are resolved into consecutive "rectangular" zones of highly concentrated pure substances rather than solvent-separated "peaks". It is primarily a preparative technique; higher product concentration, higher purity, and increased throughput may be obtained compared to other modes of chromatography.

The advent of displacement chromatography can be attributed to Arne Tiselius, who in 1943 first classified the modes of chromatography as frontal, elution, and displacement. Displacement chromatography found a variety of applications including isolation of transuranic elements and biochemical entities. The technique was redeveloped by Csaba Horváth, who employed modern high-pressure columns and equipment. It has since found many applications, particularly in the realm of biological macromolecule purification.

Principle

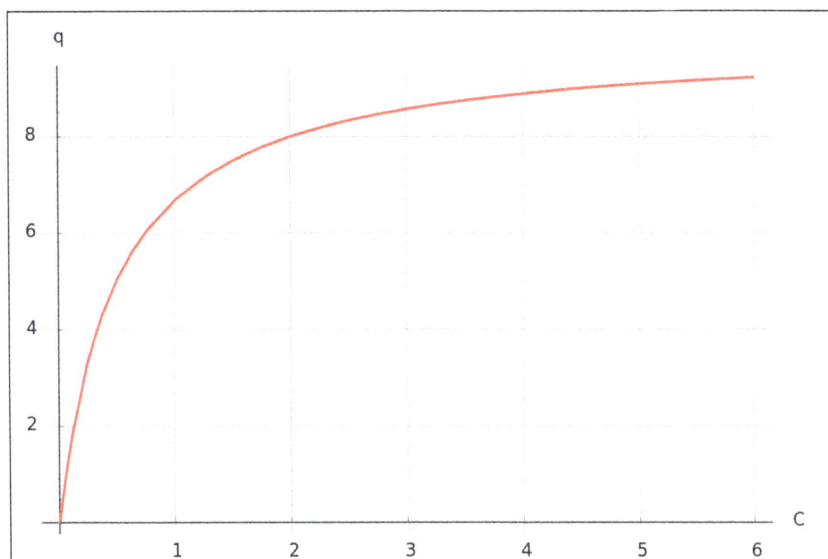

Example of a Langmuir isotherm, for a population of binding sites having uniform affinity. In this case the vertical axis represents the amount bound per unit of stationary phase, the horizontal axis the concentration in the mobile phase. In this case, the dissociation constant is 0.5 and the capacity 10; units are arbitrary.

The basic principle of displacement chromatography is: there are only a finite number of binding sites for solutes on the matrix (the stationary phase), and if a site is occupied by one molecule, it is unavailable to others. As in any chromatography, equilibrium is established between molecules of a given kind bound to the matrix and those of the same kind free in solution. Because the number of binding sites is finite, when the concentration of molecules free in solution is large relative to the dissociation constant for the sites, those sites will mostly be filled. This results in a downward-curvature in the plot of bound vs free solute, in the simplest case giving a Langmuir isotherm. A molecule with a high affinity for the matrix (the displacer) will compete more effectively for binding sites, leaving the mobile phase enriched in the lower-affinity solute. Flow of mobile phase through the column preferentially carries off the lower-affinity solute and thus at high concentration the higher-affinity solute will eventually displace all molecules with lesser affinities.

Mode of Operation

Loading

At the beginning of the run, a mixture of solutes to be separated is applied to the column, under conditions selected to promote high retention. The higher-affinity solutes are preferentially retained near the head of the column, with the lower-affinity solutes moving farther downstream. The fastest moving component begins to form a pure zone downstream. The other components also begin to form zones, but the continued supply of the mixed feed at head of the column prevents full resolution.

Displacement

After the entire sample is loaded, the feed is switched to the displacer, chosen to have higher affinity than any sample component. The displacer forms a sharp-edged zone at the head of the column, pushing the other components downstream. Each sample component now acts as a displacer for the lower-affinity solutes, and the solutes sort themselves out into a series of contiguous bands (a "displacement train"), all moving downstream at the rate set by the displacer. The size and loading of the column are chosen to let this sorting process reach completion before the components reach the bottom of the column. The solutes appear at the bottom of the column as a series of contiguous zones, each consisting of one purified component, with the concentration within each individual zone effectively uniform.

Regeneration

After the last solute has been eluted, it is necessary to strip the displacer from the column. Since the displacer was chosen for high affinity, this can pose a challenge. On reverse-phase materials, a wash with a high percentage of organic solvent may suffice. Large pH shifts are also often employed. One effective strategy is to remove the displacer by chemical reaction; for instance if hydrogen ion was used as displacer it can be removed by reaction with hydroxide, or a polyvalent metal ion can be removed by reaction with a chelating agent. For some matrices, reactive groups on the stationary phase can be titrated to temporarily eliminate the binding sites, for instance weak-acid ion exchangers or chelating resins can be converted to the protonated form. For gel-type ion exchangers, selectivity reversal at very high ionic strength can also provide a solution. Sometimes the displacer is specifically designed with a titratable functional group to shift its affinity. After the displacer is washed out, the column is washed as needed to restore it to its initial state for the next run.

Comparison with Elution Chromatography

Common Fundamentals

In any form of chromatography, the rate at which the solute moves down the column is a direct reflection of the percentage of time the solute spends in the mobile phase. To achieve separation in either elution or displacement chromatography, there must be appreciable differences in the affinity of the respective solutes for the stationary phase. Both methods rely on movement down the column to amplify the effect of small differences in distribution between the two phases. Distribution between the mobile and stationary phases is described by the binding isotherm, a plot of solute bound to (or partitioned into) the stationary phase as a function of concentration in the mobile phase. The isotherm is often linear, or approximately so, at low concentrations, but commonly curves (concave-downward) at higher concentrations as the stationary phase becomes saturated.

Characteristics of Elution Mode

In elution mode, solutes are applied to the column as narrow bands and, at low concentration, move down the column as approximately Gaussian peaks. These peaks continue to broaden as they travel, in proportion to the square root of the distance traveled. For two substances to be resolved, they must migrate down the column at sufficiently different rates to overcome the effects of band spreading. Operating at high concentration, where the isotherm is curved, is disadvantageous in elution chromatography because the rate of travel then depends on concentration, causing the peaks to spread and distort.

Retention in elution chromatography is usually controlled by adjusting the composition of the mobile phase (in terms of solvent composition, pH, ionic strength, and so forth) according to the type of stationary phase employed and the particular solutes to be separated. The mobile phase components generally have lower affinity for the stationary phase than do the solutes being separated, but are present at higher concentration and achieve their effects due to mass action. Resolution in elution chromatography is generally better when peaks are strongly retained, but conditions that give good resolution of early peaks lead to long run-times and excessive broadening of later peaks unless gradient elution is employed. Gradient equipment adds complexity and expense, particularly at large scale.

Advantages and Disadvantages of Displacement Mode

In contrast to elution chromatography, solutes separated in displacement mode form sharp-edged zones rather than spreading peaks. Zone boundaries in displacement chromatography are self-sharpening: if a molecule for some reason gets ahead of its band, it enters a zone in which it is more strongly retained, and will then run more slowly until its zone catches up. Furthermore, because displacement chromatography takes advantage of the non-linearity of the isotherms, loadings are deliberately high; more material can be separated on a given column, in a given time, with the purified components recovered at significantly higher concentrations. Retention conditions can still be adjusted, but the displacer controls the migration rate of the solutes. The displacer is selected to have higher affinity for the stationary phase than does any of the solutes being separated, and its concentration is set to approach saturation of the stationary phase and to give the desired migration rate of the concentration wave. High-retention conditions can be employed without gradient operation, because the displacer ensures removal of all solutes of interest in the designed run time.

Because of the concentrating effect of loading the column under high-retention conditions, displacement chromatography is well suited to purify components from dilute feed streams. However, it is also possible to concentrate material from a dilute stream at the head of a chromatographic column and then switch conditions to elute the adsorbed material in conventional isocratic or gradient modes. Therefore, this approach is not unique to displacement chromatography, although the higher loading capacity and less dilution allow greater concentration in displacement mode.

A disadvantage of displacement chromatography is that non-idealities always give rise to an overlap zone between each pair of components; this mixed zone must be collected separately for recycle or discard to preserve the purity of the separated materials. The strategy of adding spacer molecules to form zones between the components (sometimes termed "carrier displacement chromatography") has been investigated and can be useful when suitable, readily removable spacers are found. Another disadvantage is that the raw chromatogram, for instance a plot of absorbance or refractive index vs elution volume, can be difficult to interpret for contiguous zones, especially if the displacement train is not fully developed. Documentation and troubleshooting may require additional chemical analysis to establish the distribution of a given component. Another disadvantage is that the time required for regeneration limits throughput.

According to John C. Ford's theoretical studies indicate that at least for some systems, optimized overloaded elution chromatography offers higher throughput than displacement chromatography, though limited experimental tests suggest that displacement chromatography is superior (at least before consideration of regeneration time).

Applications

Historically, displacement chromatography was applied to preparative separations of amino acids and rare earth elements and has also been investigated for isotope separation.

Proteins

The chromatographic purification of proteins from complex mixtures can be quite challenging, particularly when the mixtures contain similarly retained proteins or when it is desired to enrich trace components in the feed. Further, column loading is often limited when high resolutions are required using traditional modes of chromatography (e.g. linear gradient, isocratic chromatography). In these cases, displacement chromatography is an efficient technique for the purification of proteins from complex mixtures at high column loadings in a variety of applications.

An important advance in the state of the art of displacement chromatography was the development of low molecular mass displacers for protein purification in ion exchange systems. This research was significant in that it represented a major departure from the conventional wisdom that large polyelectrolyte polymers are required to displace proteins in ion exchange systems.

Low molecular mass displacers have significant operational advantages as compared to large polyelectrolyte displacers. For example, if there is any overlap between the displacer and the protein of interest, these low molecular mass materials can be readily separated from the purified protein during post-displacement processing using standard size-based purification methods (e.g. size exclusion chromatography, ultrafiltration). In addition, the salt-dependent adsorption behavior of these low MW displacers greatly facilitates column regeneration. These displacers have been employed for a wide variety of high resolution separations in ion exchange systems. In addition, the utility of displacement chromatography for the purification of recombinant growth factors, antigenic vaccine proteins and antisense oligonucleotides has also been demonstrated. There are several examples in which displacement chromatography has been applied to the purification of proteins using ion exchange, hydrophobic interaction, as well as reversed-phase chromatography.

Displacement chromatography is well suited for obtaining mg quantities of purified proteins from complex mixtures using standard analytical chromatography columns at the bench scale.

It is also particularly well suited for enriching trace components in the feed. Displacement chromatography can be readily carried out using a variety of resin systems including, ion exchange, HIC and RPLC.

Two-dimensional Chromatography

Two-dimensional chromatography represents the most thorough and rigorous approach to evaluation of the proteome. While previously accepted approaches have utilized elution mode chromatographic approaches such as cation exchange to reversed phase HPLC, yields are typically very low requiring analytical sensitivities in the picomolar to femtomolar range. As displacement chromatography offers the advantage of concentration of trace components, two dimensional chromatography utilizing displacement rather than elution mode in the upstream chromatography step represents a potentially powerful tool for analysis of trace components, modifications, and identification of minor expressed components of the proteome.

Size-exclusion Chromatography

Size-exclusion chromatography (SEC), also known as molecular sieve chromatography, is a chromatographic method in which molecules in solution are separated by their size, and in some cases molecular weight. It is usually applied to large molecules or macromolecular complexes such as proteins and industrial polymers. Typically, when an aqueous solution is used to transport the sample through the column, the technique is known as gel-filtration chromatography, versus the name gel permeation chromatography, which is used when an organic solvent is used as a mobile phase. The chromatography column is packed with fine, porous beads which are composed of dextran polymers (Sephadex), agarose (Sepharose), or polyacrylamide (Sephacryl or BioGel P). The pore sizes of these beads are used to estimate the dimensions of macromolecules. SEC is a widely used polymer characterization method because of its ability to provide good molar mass distribution (Mw) results for polymers.

Applications

The main application of gel-filtration chromatography is the fractionation of proteins and other water-soluble polymers, while gel permeation chromatography is used to analyze the molecular weight distribution of organic-soluble polymers. Either technique should not be confused with gel electrophoresis, where an electric field is used to "pull" or "push" molecules through the gel depending on their electrical charges. The amount of time a solute remains within a pore is dependent on the size of the pore. Larger solutes will have access to a smaller volume and vice versa. Therefore, a smaller solute will remain within the pore for a longer period of time compared to a larger solute.

Another use of size exclusion chromatography is to examine the stability and characteristics of natural organic matter in water. In this method, Margit B. Muller, Daniel Schmitt, and Fritz H. Frimmel tested water sources from different places in the world to determine how stable the natural organic matter is over a period of time. Even though, size exclusion chromatography is widely utilized to study natural organic material, there are limitations. One of these limitations include that there is no standard molecular weight marker; thus, there is nothing to compare the results back to. If precise molecular weight is required, other methods should be used.

Advantages

The advantages of this method include good separation of large molecules from the small molecules with a minimal volume of eluate, and that various solutions can be applied without interfering with the filtration process, all while preserving the biological activity of the particles to separate. The technique is generally combined with others that further separate molecules by other characteristics, such as acidity, basicity, charge, and affinity for certain compounds. With size exclusion chromatography, there are short and well-defined separation times and narrow bands, which lead to good sensitivity. There is also no sample loss because solutes do not interact with the stationary phase.

The other advantage to this experimental method is that in certain cases, it is feasible to determine the approximate molecular weight of a compound. The shape and size of the compound (eluent) determine how the compound interacts with the gel (stationary phase). To determine approximate molecular weight, the elution volumes of compounds with their corresponding molecular weights the elution volumes of compounds with their corresponding molecular weights are obtained and then a plot of "K_{av}" vs "log(Mw)" is made, where $K_{av} = (V_e - V_o) / (V_t - V_o)$ and Mw is the molecular mass. This plot acts as a calibration curve, which is used to approximate the desired compound's molecular weight. The V_e component represents the volume at which the intermediate molecules elute such as molecules that have partial access to the beads of the column. In addition, V_t is the sum of the total volume between the beads and the volume within the beads. The V_o component represents the volume at which the larger molecules elute, which elute in the beginning. Disadvantages are, for example, that only a limited number of bands can be accommodated because the time scale of the chromatogram is short, and, in general, there must be a 10% difference in molecular mass to have a good resolution.

The technique was invented by Grant Henry Lathe and Colin R Ruthven, working at Queen Charlotte's Hospital, London. They later received the John Scott Award for this invention. While Lathe and Ruthven used starch gels as the matrix, Jerker Porath and Per Flodin later introduced dextran gels; other gels with size fractionation properties include agarose and polyacrylamide. A short review of these developments has appeared.

There were also attempts to fractionate synthetic high polymers; however, it was not until 1964, when J. C. Moore of the Dow Chemical Company published his work on the preparation of gel permeation chromatography (GPC) columns based on cross-linked polystyrene with controlled pore size, that a rapid increase of research activity in this field began. It was recognized almost immediately that with proper calibration, GPC was capable to provide molar mass and molar mass distribution information for synthetic polymers. Because the latter information was difficult to obtain by other methods, GPC came rapidly into extensive use.

SEC is used primarily for the analysis of large molecules such as proteins or polymers. SEC works by trapping smaller molecules in the pores of the adsorbent ("stationary phase"). This process is usually performed within a column, which typically consists of a hollow tube tightly packed with micron-scale polymer beads containing pores of different sizes. These pores may be depressions on the surface or channels through the bead. As the solution travels down the column some particles enter into the pores. Larger particles cannot enter into as many pores. The larger the particles, the faster the elution. The larger molecules simply pass by the pores because those molecules are too large to enter the pores. Larger molecules therefore flow through the column more quickly than smaller molecules, that is, the smaller the molecule, the longer the retention time.

One requirement for SEC is that the analyte does not interact with the surface of the stationary phases, with differences in elution time between analytes ideally being based solely on the solute volume the analytes can enter, rather than chemical or electrostatic interactions with the stationary phases. Thus, a small molecule that can penetrate every region of the stationary phase pore system can enter a total volume equal to the sum of the entire pore volume and the interparticle volume. This small molecule elutes late (after the molecule has penetrated all of the pore- and interparticle volume—approximately 80% of the column volume). At the other extreme, a very large molecule that cannot penetrate any the smaller pores can enter only the interparticle volume (~35% of the column volume) and elutes earlier when this volume of mobile phase has passed through the column. The underlying principle of SEC is that particles of different sizes elute (filter) through a stationary phase at different rates. This results in the separation of a solution of particles based on size. Provided that all the particles are loaded simultaneously or near-simultaneously, particles of the same size should elute together.

However, as there are various measures of the size of a macromolecule (for instance, the radius of gyration and the hydrodynamic radius), a fundamental problem in the theory of SEC has been the choice of a proper molecular size parameter by which molecules of different kinds are separated. Experimentally, Benoit and co-workers found an excellent correlation between elution volume and a dynamically based molecular size, the hydrodynamic volume, for several different chain architecture and chemical compositions. The observed correlation based on the hydrodynamic volume became accepted as the basis of universal SEC calibration.

Still, the use of the hydrodynamic volume, a size based on dynamical properties, in the interpretation of SEC data is not fully understood. This is because SEC is typically run under low flow rate conditions where hydrodynamic factor should have little effect on the separation. In fact, both theory and computer simulations assume a thermodynamic separation principle: the separation process is determined by the equilibrium distribution (partitioning) of solute macromolecules between two phases — a dilute bulk solution phase located at the interstitial space and confined solution phases within the pores of column packing material. Based on this theory, it has been shown that the relevant size parameter to the partitioning of polymers in pores is the mean span dimension (mean maximal projection onto a line). Although this issue has not been fully resolved, it is likely that the mean span dimension and the hydrodynamic volume are strongly correlated.

Each size exclusion column has a range of molecular weights that can be separated. The exclusion limit defines the molecular weight at the upper end of the column 'working' range and is where molecules are too large to get trapped in the stationary phase. The lower end of the range is defined by the permeation limit, which defines the molecular weight of a molecule that is small enough to penetrate all pores of the stationary phase. All molecules below this molecular mass are so small that they elute as a single band.

The filtered solution that is collected at the end is known as the eluate. The void volume includes any particles too large to enter the medium, and the solvent volume is known as the column volume.

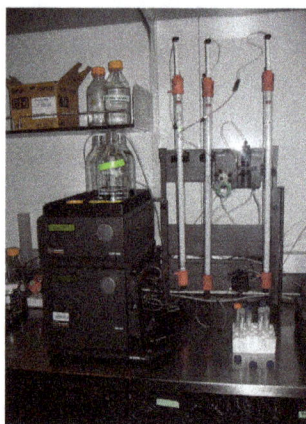

Agarose-based SEC columns used for protein
purification on an AKTA FPLC machine.

A size exclusion column.

Factors affecting Filtration

In real-life situations, particles in solution do not have a fixed size, resulting in the probability that a particle that would otherwise be hampered by a pore passing right by it. Also, the stationary-phase particles are not ideally defined; both particles and pores may vary in size. Elution curves, therefore, resemble Gaussian distributions. The stationary phase may also interact in undesirable ways with a particle and influence retention times, though great care is taken by column manufacturers to use stationary phases that are inert and minimize this issue.

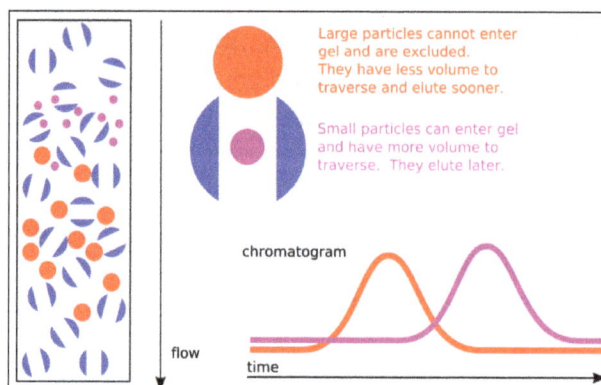

Large particles cannot enter gel and are excluded. They have less volume to traverse and elute sooner.

Small particles can enter gel and have more volume to traverse. They elute later.

chromatogram

flow

time

The theory behind size exclusion chromatography.

Like other forms of chromatography, increasing the column length enhances resolution, and increasing the column diameter increases column capacity. Proper column packing is important for maximum resolution: An over-packed column can collapse the pores in the beads, resulting in a loss of resolution. An under-packed column can reduce the relative surface area of the stationary phase accessible to smaller species, resulting in those species spending less time trapped in pores. Unlike affinity chromatography techniques, a solvent head at the top of the column can drastically diminish resolution as the sample diffuses prior to loading, broadening the downstream elution.

Analysis

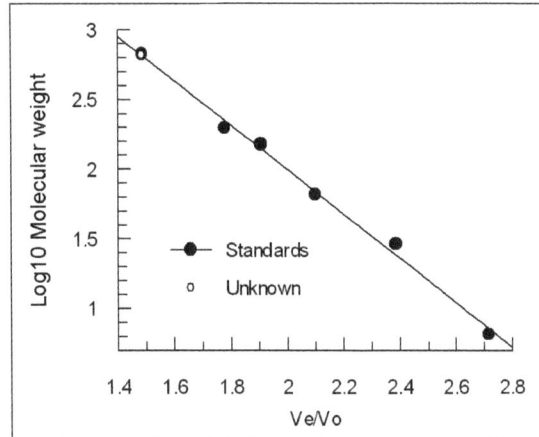

Standardization of a size exclusion column.

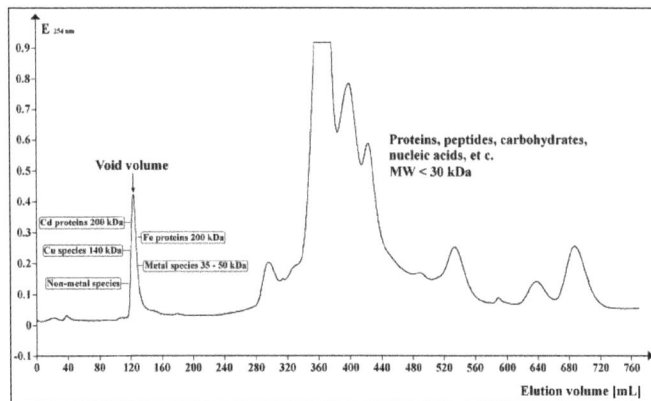

SEC Chromatogram of a biological sample.

In simple manual columns, the eluent is collected in constant volumes, known as fractions. The more similar the particles are in size the more likely they are in the same fraction and not detected separately. More advanced columns overcome this problem by constantly monitoring the eluent.

The collected fractions are often examined by spectroscopic techniques to determine the concentration of the particles eluted. Common spectroscopy detection techniques are refractive index (RI) and ultraviolet (UV). When eluting spectroscopically similar species (such as during biological purification), other techniques may be necessary to identify the contents of each fraction. It is also possible to analyse the eluent flow continuously with RI, LALLS, Multi-Angle Laser Light Scattering MALS, UV, and/or viscosity measurements.

The elution volume (V_e) decreases roughly linear with the logarithm of the molecular hydrodynamic volume. Columns are often calibrated using 4-5 standard samples (e.g., folded proteins of known molecular weight), and a sample containing a very large molecule such as thyroglobulin to determine the void volume. (Blue dextran is not recommended for V_o determination because it is heterogeneous and may give variable results) The elution volumes of the standards are divided by the elution volume of the thyroglobulin (V_e/V_o) and plotted against the log of the standards' molecular weights.

Applications

Biochemical Applications

In general, SEC is considered a low resolution chromatography as it does not discern similar species very well, and is therefore often reserved for the final step of a purification. The technique can determine the quaternary structure of purified proteins that have slow exchange times, since it can be carried out under native solution conditions, preserving macromolecular interactions. SEC can also assay protein tertiary structure, as it measures the hydrodynamic volume (not molecular weight), allowing folded and unfolded versions of the same protein to be distinguished. For example, the apparent hydrodynamic radius of a typical protein domain might be 14 Å and 36 Å for the folded and unfolded forms, respectively. SEC allows the separation of these two forms, as the folded form elutes much later due to its smaller size.

Polymer Synthesis

SEC can be used as a measure of both the size and the polydispersity of a synthesised polymer, that is, the ability to find the distribution of the sizes of polymer molecules. If standards of a known size are run previously, then a calibration curve can be created to determine the sizes of polymer molecules of interest in the solvent chosen for analysis (often THF). In alternative fashion, techniques such as light scattering and/or viscometry can be used online with SEC to yield absolute molecular weights that do not rely on calibration with standards of known molecular weight. Due to the difference in size of two polymers with identical molecular weights, the absolute determination methods are, in general, more desirable. A typical SEC system can quickly (in about half an hour) give polymer chemists information on the size and polydispersity of the sample. The preparative SEC can be used for polymer fractionation on an analytical scale.

Drawback

In SEC, mass is not measured so much as the hydrodynamic volume of the polymer molecules, that is, how much space a particular polymer molecule takes up when it is in solution. However, the approximate molecular weight can be calculated from SEC data because the exact relationship between molecular weight and hydrodynamic volume for polystyrene can be found. For this, polystyrene is used as a standard. But the relationship between hydrodynamic volume and molecular weight is not the same for all polymers, so only an approximate measurement can be obtained. Another drawback is the possibility of interaction between the stationary phase and the analyte. Any interaction leads to a later elution time and thus mimics a smaller analyte size.

When performing this method, the bands of the eluting molecules may be broadened. This can occur by turbulence caused by the flow of the mobile phase molecules passing through the molecules of the stationary phase. In addition, molecular thermal diffusion and friction between the molecules of the glass walls and the molecules of the eluent contribute to the broadening of the bands. Besides broadening, the bands also overlap with each other. As a result, the eluent usually gets considerably diluted. A few precautions can be taken to prevent the likelihood of the bands broadening. For instance, one can apply the sample in a narrow, highly concentrated band on the top of the column. The more concentrated the eluent is, the more efficient the procedure would be. However, it is not always possible to concentrate the eluent, which can be considered as one more disadvantage.

Absolute Size-exclusion Chromatography

Absolute size-exclusion chromatography (ASEC) is a technique that couples a dynamic light scattering (DLS) instrument to a size exclusion chromatography system for absolute size measurements of proteins and macromolecules as they elute from the chromatography system.

The definition of "absolute" in this case is that calibration is not required to obtain hydrodynamic size, often referred to as hydrodynamic diameter (DH in units of nm). The sizes of the macromolecules are measured as they elute into the flow cell of the DLS instrument from the size exclusion column set. The hydrodynamic size of the molecules or particles are measured and not their molecular weights. For proteins a Mark-Houwink type of calculation can be used to estimate the molecular weight from the hydrodynamic size.

A major advantage of DLS coupled with SEC is the ability to obtain enhanced DLS resolution. Batch DLS is quick and simple and provides a direct measure of the average size, but the baseline resolution of DLS is 3 to 1 in diameter. Using SEC, the proteins and protein oligomers are separated, allowing oligomeric resolution. Aggregation studies can also be done using ASEC. Though the aggregate concentration may not be calculated, the size of the aggregate can be measured, only limited by the maximum size eluting from the SEC columns.

Limitations of ASEC include flow-rate, concentration, and precision. Because a correlation function requires anywhere from 3–7 seconds to properly build, a limited number of data points can be collected across the peak.

Micellar Electrokinetic Chromatography

Micellar electrokinetic chromatography (MEKC) is a chromatography technique used in analytical chemistry. It is a modification of capillary electrophoresis (CE), extending its functionality to neutral analytes, where the samples are separated by differential partitioning between micelles (pseudo-stationary phase) and a surrounding aqueous buffer solution (mobile phase).

The basic set-up and detection methods used for MEKC are the same as those used in CE. The difference is that the solution contains a surfactant at a concentration that is greater than the critical micelle concentration (CMC). Above this concentration, surfactant monomers are in equilibrium with micelles.

In most applications, MEKC is performed in open capillaries under alkaline conditions to generate a strong electroosmotic flow. Sodium dodecyl sulfate (SDS) is the most commonly used surfactant in MEKC applications. The anionic character of the sulfate groups of SDS cause the surfactant and micelles to have electrophoretic mobility that is counter to the direction of the strong electroosmotic flow. As a result, the surfactant monomers and micelles migrate quite slowly, though their net movement is still toward the cathode. During a MEKC separation, analytes distribute themselves between the hydrophobic interior of the micelle and hydrophilic buffer solution.

Analytes that are insoluble in the interior of micelles should migrate at the electroosmotic flow velocity, u_o, and be detected at the retention time of the buffer, t_M. Analytes that solubilize completely within the micelles (analytes that are highly hydrophobic) should migrate at the micelle velocity, u_c, and elute at the final elution time, t_c.

The micelle velocity is defined by:

$$u_c = u_p + u_o$$

where u_p is the electrophoretic velocity of a micelle. The retention time of a given sample should depend on the capacity factor, k^1:

$$k^1 = \frac{n_c}{n_w}$$

where n_c is the total number of moles of solute in the micelle and n_w is the total moles in the aqueous phase. The retention time of a solute should then be within the range:

$$t_M \leq t_r \leq t_c$$

Charged analytes have a more complex interaction in the capillary because they exhibit electrophoretic mobility, engage in electrostatic interactions with the micelle, and participate in hydrophobic partitioning.

The fraction of the sample in the aqueous phase, R, is given by:

$$R = \frac{u_s - u_c}{u_o - u_c}$$

where, u_s is the migration velocity of the solute. The value R can also be expressed in terms of the capacity factor:

$$R = \frac{1}{1 + k^1}$$

Using the relationship between velocity, tube length from the injection end to the detector cell (L), and retention time, $u_o = L / t_M$, $u_c = L / t_c$, and $u_s = L / t_r$, a relationship between the capacity factor and retention times can be formulated:

$$k^1 = \frac{t_r - t_M}{t_M (1 - (t_r / t_c))}$$

The extra term enclosed in parenthesis accounts for the partial mobility of the hydrophobic phase in MEKC. This equation resembles an expression derived for k^1 in conventional packed bed chromatography:

$$k = \frac{t_r - t_M}{t_M}$$

A rearrangement of the previous equation can be used to write an expression for the retention factor:

$$t_r = \left(\frac{1 + k^1}{1 + (t_M / t_c)k^1} \right) t_M$$

From this equation it can be seen that all analytes that partition strongly into the micellar phase (where k^1 is essentially ∞) migrate at the same time, t_c. In conventional chromatography, separation of similar compounds can be improved by gradient elution. In MEKC, however, techniques must be used to extend the elution range to separate strongly retained analytes.

Elution ranges can be extended by several techniques including the use of organic modifiers, cyclodextrins, and mixed micelle systems. Short-chain alcohols or acetonitrile can be used as organic modifiers that decrease t_M and k^1 to improve the resolution of analytes that co-elute with the micellar phase. These agents, however, may alter the level of the EOF. Cyclodextrins are cyclic polysaccharides that form inclusion complexes that can cause competitive hydrophobic partitioning of the analyte. Since analyte-cyclodextrin complexes are neutral, they will migrate toward the cathode at a higher velocity than that of the negatively charged micelles. Mixed micelle systems, such as the one formed by combining SDS with the neutral surfactant Brij-35, can also be used to alter the selectivity of MEKC.

Applications

The simplicity and efficiency of MEKC have made it an attractive technique for a variety of applications. Further improvements can be made to the selectivity of MEKC by adding chiral selectors or chiral surfactants to the system. Unfortunately, this technique is not suitable for protein analysis because proteins are generally too large to partition into a surfactant micelle and tend to bind to surfactant monomers to form SDS-protein complexes.

Recent applications of MEKC include the analysis of uncharged pesticides, essential and branched-chain amino acids in nutraceutical products, hydrocarbon and alcohol contents of the marjoram herb.

MEKC has also been targeted for its potential to be used in combinatorial chemical analysis. The advent of combinatorial chemistry has enabled medicinal chemists to synthesize and identify large numbers of potential drugs in relatively short periods of time. Small sample and solvent requirements and the high resolving power of MEKC have enabled this technique to be used to quickly analyze a large number of compounds with good resolution.

Traditional methods of analysis, like high-performance liquid chromatography (HPLC), can be used to identify the purity of a combinatorial library, but assays need to be rapid with good resolution for all components to provide useful information for the chemist. The introduction of surfactant to traditional capillary electrophoresis instrumentation has dramatically expanded the scope of analytes that can be separated by capillary electrophoresis.

MEKC can also be used in routine quality control of antibiotics in pharmaceuticals or feedstuffs.

Two-dimensional Chromatography

Two-dimensional chromatography is a type of chromatographic technique in which the injected sample is separated by passing through two different separation stages. Two different chromatographic columns are connected in sequence, and the effluent from the first system is transferred onto the second column. Typically the second column has a different separation mechanism, so that bands that are poorly resolved from the first column may be completely separated in the second column. (For instance, a C18 reversed-phase chromatography column may be followed by a phenyl column.) Alternately, the two columns might run at different temperatures. During the second stage of separation the rate at which the separation occurs must be faster than the first stage, since there is still only a single detector. The plane surface is amenable to sequential development in two directions using two different solvents.

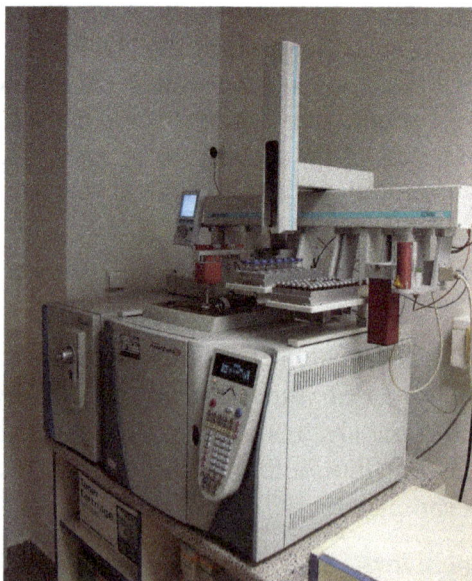

Examples:

Two-dimensional separations can be carried out in gas chromatography or liquid chromatography. Various different coupling strategies have been developed to "resample" from the first column into the second. Some important hardware for two-dimensional separations are Deans' switch and Modulator, which selectively transfer the first dimension eluent to second dimension column.

The chief advantage of two-dimensional techniques is that they offer a large increase in peak capacity, without requiring extremely efficient separations in either column. (For instance, if the first column offers a peak capacity (k_1)of 100 for a 10-minute separation, and the second column offers a peak capacity of 5 (k_2) in a 5-second separation, then the combined peak capacity may approach $k_1 \times k_2 = 500$, with the total separation time still ~ 10 minutes). 2D separations have been applied to the analysis of gasoline and other petroleum mixtures, and more recently to protein mixtures.

Tandem Mass Spectrometry

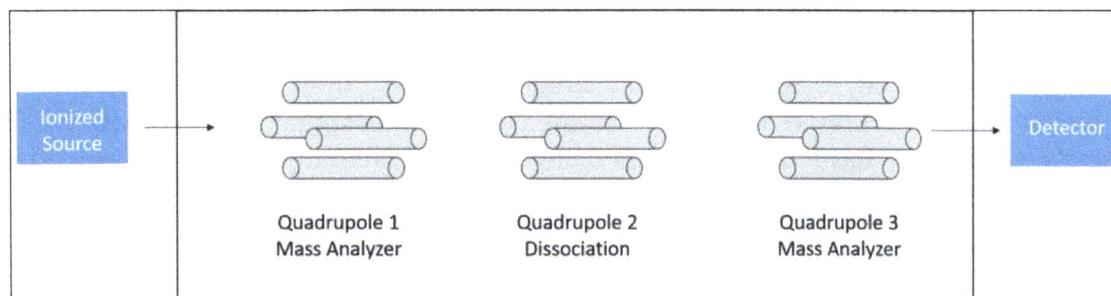

A triple quadrupole (QQQ) mass analyzer.

Tandem mass spectrometry (Tandem MS or MS/MS) uses two mass analyzers in sequence to separate more complex mixtures of analytes. The advantage of tandem MS is that it can be much faster than other two-dimensional methods, with times ranging from milliseconds to seconds. Because there is no dilution with solvents in MS, there is less probability of interference, so tandem MS can be more sensitive and have a higher signal-to-noise ratio compared to other two-dimensional methods. The main disadvantage associated with tandem MS is the high cost of the instrumentation needed. Prices can range from $500,000 to over $1 million. Many form of tandem MS involve a mass selection step and a fragmentation step. The first mass analyzer can be programmed to only pass molecules of a specific mass-to-charge ratio. Then the second mass analyzer can fragment the molecule to determine its identity. This can be especially useful for separating molecules of the same mass (i.e. proteins of the same mass or molecular isomers). Different types of mass analyzers can be coupled to achieve varying effects. One example would be a TOF-Quadrople system. Ions can be sequentially fragmented and/or analyzed in a quadrupole as they leave the TOF in order of increasing m/z. Another prevalent tandem mass spectrometer is the quardupole-quadrupole-quadrupole (Q-Q-Q) analyzer. The first quadrupole separates by mass, collisions take place in the second quadruple, and the fragments are separated by mass in the third quadrupole.

Tandem MS has been gaining popularity and relevance as analytical techniques are becoming more and more precise. This is an active area of research by many analytical chemists around the world, including David E. Clemmer of Indiana University, who is in the forefront of mass spectrometer instrumentation.

Gas Chromatography-Mass Spectrometry

Gas Chromatography-Mass Spectrometry (GC-MS) is a two-dimensional chromatography technique that combines the separation technique of gas chromatography with the identification technique of mass spectrometry. GC-MS is the single most important analytical tool for the analysis of volatile and semi-volatile organic compounds in complex mixtures. It works by first injecting the sample into the GC inlet where it is vaporized and pushed through a column by a carrier gas, typically helium. The analytes in the sample are separated based upon their interaction with the coating of the column, or the stationary phase, and the carrier gas, or the mobile phase. The compounds eluted from the column are converted into ions via electron impact (EI) or chemical ionization (CI) before traveling through the mass analyzer. The mass analyzer serves to separate the ions on a mass-to-charge basis. Popular choices perform the same function but differ in the way that they accomplish the separation. The analyzers typically used with GC-MS are the time-of-flight mass analyzer and the quadrupole mass analyzer. After leaving the mass analyzer, the analytes reach the detector and produce a signal that is read by a computer and used to create a gas chromatogram and mass spectrum. Sometimes GC-MS utilizes two gas chromatographers in particularly complex samples to obtain considerable separation power and be able to unambiguously assign the specific species to the appropriate peaks in a technique known as GCxGC-(MS). Ultimately, GC-MS is a technique utilized in many analytical laboratories and is a very effective and adaptable analytical tool.

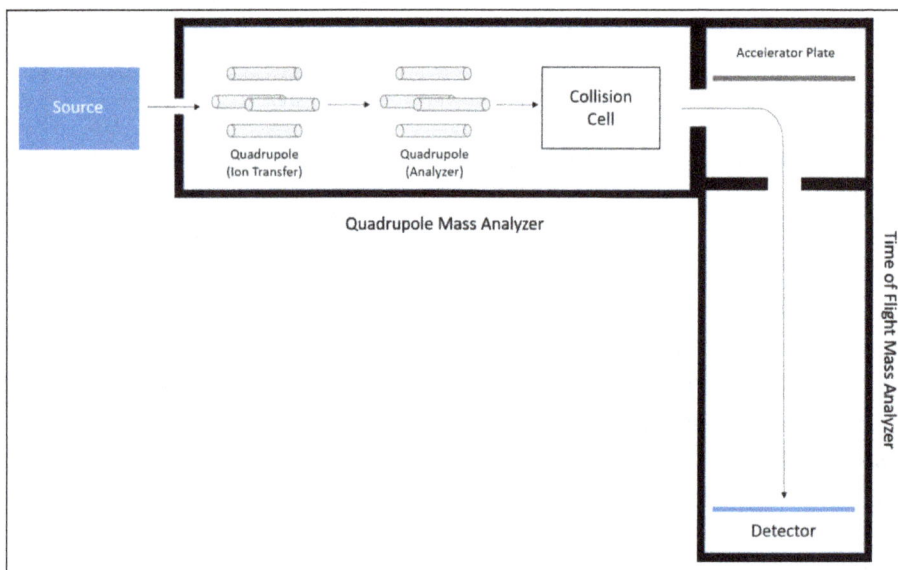

A quadrupole/time of flight mass analyzer. The ionized source sample first enters the quadrupole mass analyzer, then enters the time of flight analyzer.

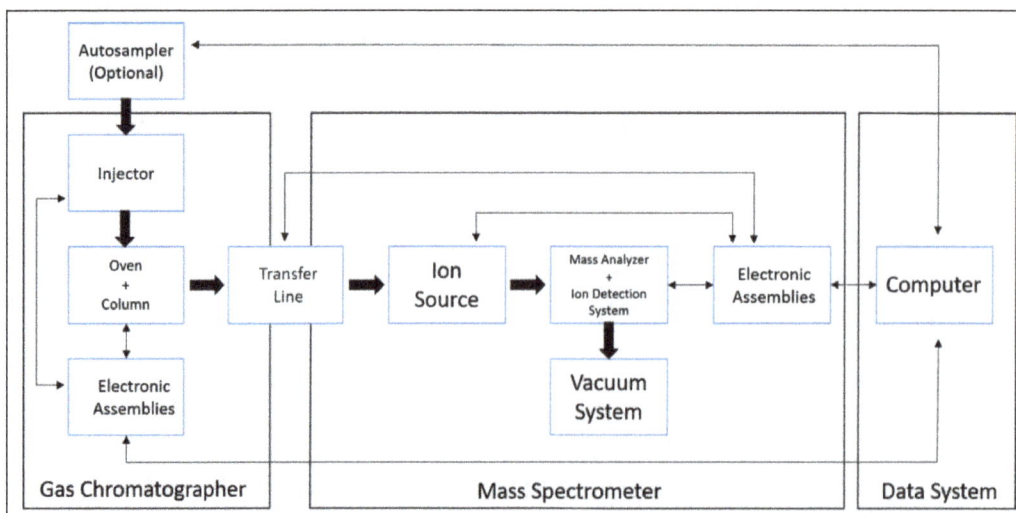

The diagram shows the pathway of the analyte. The analyte first passes through the gas chromatographer and then the separated analytes are subjected to mass analysis. Different types of mass analyzers, ToF, qudrupole, etc., can by used in the MS.

Liquid Chromatography-Mass Spectrometry

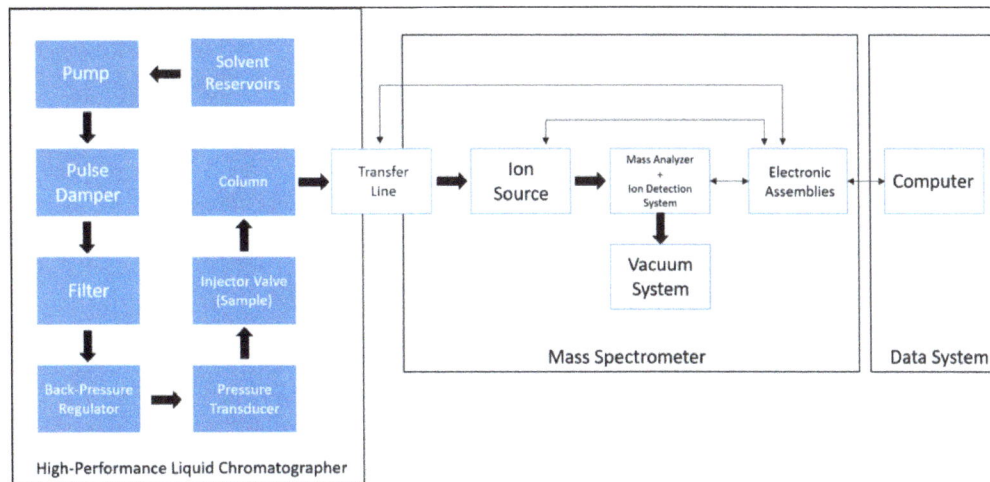

The sample is first subjected to analsis by HPLC and then is subjected to mass analysis.
Different types of mass analyzers, ToF, qudrupole, etc., can be used in the MS.

Liquid Chromatography-Mass Spectrometry (LC/MS) couples high resolution chromatographic separation with MS detection. As the system adopts the high separation of HPLC, analytes which are in the liquid mobile phase are often ionized by various soft ionization methods including atmospheric pressure chemical ionization (APCI), electrospray ionization (ESI) or matrix-assisted laser desorption/ionization (MALDI), which attains the gas phase ionization required for the coupling with MS. These ionization methods allow the analysis of a wider range of biological molecules, including those with larger masses, thermally unstable or nonvolatile compounds where GC-MS is typically incapable of analyzing.

LC-MS provides high selectivity as unresolved peaks can be isolated by selecting a specific mass. Furthermore, better identification is also attained by mass spectra and the user does not have to rely solely on the retention time of analytes. As a result, molecular mass and structural information as well as quantitative data can all be obtained via LC-MS. LC-MS can therefore be applied to various fields, such as impurity identification and profiling in drug development and pharmaceutical manufacturing, since LC provides efficient separation of impurities and MS provides structural characterization for impurity profiling.

Common solvents used in normal or reversed phase LC such as water, acetonitrile, and methanol are all compatible with ESI, yet a LC grade solvent may not be suitable for MS. Furthermore, buffers containing inorganic ions should be avoided as they may contaminate the ion source. Nonetheless, the problem can be resolved by 2D LC-MS, as well as other various issues including analyte coelusion and UV detection responses.

Liquid Chromatography-Liquid Chromatography

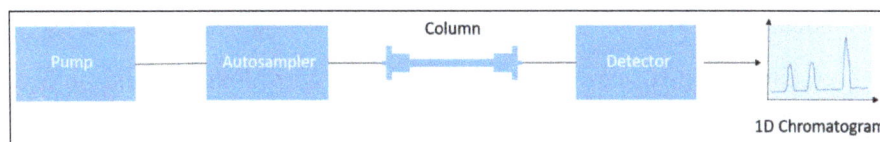

1-Dimensional LC.

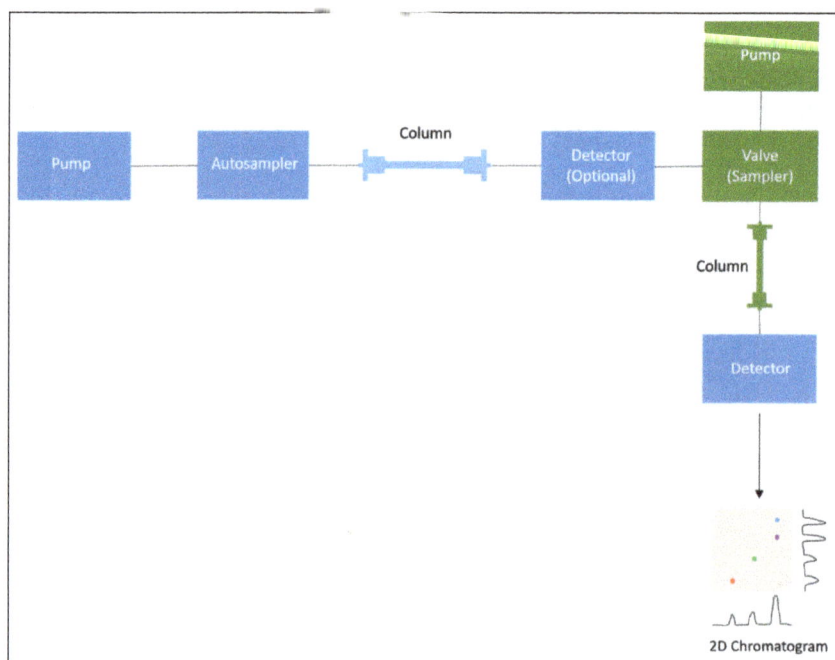

2-Dimensional LC.

Two dimensional liquid chromatography (2D-LC) combines two separate analyses of liquid chromatography into one data analysis. Modern 2-D liquid chromatography has its origins in the late 1970s to early 1980's. During this time, the hypothesized principles of 2D-LC were being proven via experiments conducted along with supplementary conceptual and theoretical work. It was shown that 2D-LC could offer quite a bit more resolving power compared to the conventional techniques of one-dimensional liquid chromatography. In the 1990s, the technique of 2D-LC played an important role in the separation of extremely complex substances and materials found in the proteomics and polymer fields of study. Unfortunately, the technique had been shown to have a significant disadvantage when it came to analysis time. Early work with 2D-LC was limited to small portion of liquid phase separations due to the long analysis time of the machinery. Modern 2D-LC techniques tackled that disadvantage head on, and have significantly reduced what was once a damaging feature. Modern 2D-LC has an instrumental capacity for high resolution separations to be completed in an hour or less. Due to the growing need for instrumentation to perform analysis on substances of growing complexity with better detection limits, the development of 2D-LC pushes forward. Instrumental parts have become a mainstream industry focus and are much easier to attain then before. Prior to this, 2D-LC was performed using components from 1D-LC instruments, and would lead to results of varying degrees in both accuracy and precision. The reduced stress on instrumental engineering has allowed for pioneering work in the field and technique of 2D-LC.

The purpose of employing this technique is to separate mixtures that one dimensional liquid chromatography otherwise cannot separate effectively. Two dimensional liquid chromatography is better suited to analyzing complex mixtures samples such as urine, environmental substances and forensic evidence such as blood.

Difficulties in separating mixtures can be attributed to the complexity of the mixture in the sense that separation cannot occur due to the number of different effluents in the compound. Another problem associated with one dimensional liquid chromatography involves the difficulty associated to resolving closely related compounds. Closely related compounds have similar chemical properties that may prove difficult to separate based on polarity, charge, etc. Two dimensional liquid chromatography provides separation based on more than one chemical or physical property. Using an example from Nagy and Vekey, a mixture of peptides can be separated based on their basicity, but similar peptides may not elute well. Using a subsequent LC technique, the similar basicity between the peptides can be further separated by employing differences in apolar character.

As a result, to be able to separate mixtures more efficiently, a subsequent LC analysis must employ very different separation selectivity relative to the first column. Another requirement to effectively use 2D liquid chromatography, according to Bushey and Jorgenson, is to employ highly orthogonal techniques which means that the two separation techniques must be as different as possible.

There are two major classifications of 2D liquid chromatography. These include: Comprehensive 2D liquid chromatography (LCxLC) and Heart-cutting 2D liquid chromatrography (LC-LC). In comprehensive 2D-LC, all the peaks from a column elution are fully sampled, but it has been deemed unnecessary to transfer the entire sample from the first to the second column. A portion of the sample is sent to waste while the rest is sent to the sampling valve. In heart-cutting 2D-LC specific peaks are targeted with only a small portion of the peak being injected onto a second column. Heart-cutting 2D-LC has proven to be quite useful for sample analysis of substances that are not very complex provided they have similar retention behavior. Compared to comprehensive 2D-LC, heart-cutting 2D-LC provides an effective technique with much less system setup and a much lower operating cost. Multiple heart-cutting (mLC-LC) may be utilized to sample multiple peaks from first dimensional analysis without risking temporary overlap of second dimensional analysis. Multiple heart-cutting (mLC-LC) utilizes a setup of multiple sampling loops.

For 2D-LC, peak capacity is a very important issue. This can be generated using Gradient elution separation with much greater efficiency than an isocratic separation given a reasonable amount of time. While isocratic elution is much easier on a fast time scale, it is preferable to perform a gradient elution separation in the second dimension. The mobile phase strength is varied from a weak eluent composition to a stronger one. Based on linear solvent strength theory (LSST) of gradient elution for reversed phase chromatography, the relationship between retention time, instrumental variables and solute parameters is shown.

$$t_R = t_o + t_D + t_o/b * \ln(b * (k_o - t_d/t_o) + 1)$$

While a lot of pioneering work has been completed in the years since 2D-LC became a major analytical chromatographic technique, there are still many modern problems to be considered. Large amounts of experimental variables have yet to be decided on, and the technique is constantly in a state of development.

Gas Chromatography-Gas Chromatography

Comprehensive two-dimensional gas chromatography is an analytical technique that separates and analyzes complex mixtures. It has been utilized in fields such as: flavor, fragrance, environmental studies, pharmaceuticals, petroleum products and forensic science. GCxGC provides a high range of sensitivity and produces a greater separation power due to the increased peak capacity.

Affinity Chromatography

Affinity chromatography is a method of separating biochemical mixture based on a highly specific interaction between antigen and antibody, enzyme and substrate, receptor and ligand, or protein and nucleic acid. It is a type of chromatographic laboratory technique used for purifying biological molecules within a mixture by exploiting molecular properties.

Biological macromolecules, such as enzymes and other proteins, interact with other molecules with high specificity through several different types of bonds and interaction. Such interactions include hydrogen bonding, ionic interaction, disulfide bridges, hydrophobic interaction, and more. The high selectivity of affinity chromatography is caused by allowing the desired molecule to interact with the stationary phase and be bound within the column in order to be separated from the undesired material which will not interact and elute first. The molecules no longer needed are first washed away with a buffer while the desired proteins are let go in the presence of the eluting solvent (of higher salt concentration). This process creates a competitive interaction between the desired protein and the immobilized stationary molecules, which eventually lets the now highly purified proteins be released.

Uses

Affinity chromatography can be used to purify and concentrate a substance from a mixture into a buffering solution, reduce the amount of unwanted substances in a mixture, identify the biological compounds binding to a particular substance, purify and concentrate an enzyme solution. The molecule of interest can be immobilized through covalent bonds. This occurs through an insoluble matrix such as chromatographic medium like cellulose or polyacrylamide. When the medium is bound to the protein of interest it becomes immobilized.

Immunochromatographic Test

Affinity chromatography is the basis for immunochromatographic test (ICT) strips, which provide a rapid means of diagnosis in patient care. Using ICT, a technician can make a determination at a patient's bedside, without the need for a laboratory. ICT detection is highly specific to the microbe causing an infection.

Principle

Affinity chromatography exploits the differences in interactions' strengths between the different biomolecules within a mobile phase, and the stationary phase. The stationary phase is first loaded into a column with mobile phase containing a variety of biomolecules from DNA to proteins (depending on the purification experiment). Then, the two phases are allowed time to bind. A wash buffer is then poured through a column containing both bound phases. The wash buffer removes non-target biomolecules by disrupting their weaker interactions with the stationary phase. Target biomolecules have a much higher affinity for the stationary phase, and remain bound to the stationary phase, not being washed away by wash buffer. An elution buffer is then poured through the column containing the remaining target biomolecules. The elution buffer disrupts interactions between the bound target biomolecules with the stationary to a much greater extent than the wash buffer, effectively removing the target biomolecules. This purified solution contains elution buffer and target biomolecules, and is called elution.

The stationary phase is typically a gel matrix, often of agarose; a linear sugar molecule derived from algae. To prevent steric interference or overlap during the binding process of the target molecule to the ligand, an inhibitor containing a hydrocarbon chain is first attached to the agarose bead (solid support). This inhibitor with a hydrocarbon chain is commonly known as the spacer between the agarose bead and the target molecule.

Usually, the starting point is a crude, heterogeneous group of molecules in a whole cell extract, such as a cell lysate, growth medium or blood serum. The molecule of interest will have a well known and defined property, and can be exploited during the affinity purification process. The process itself can be thought of as an entrapment, with the target molecule becoming trapped on a solid or stationary phase or medium. The other molecules in the mobile phase will not become trapped as they do not possess this property. The stationary phase can then be removed from the mixture, washed and the target molecule released from the entrapment in a process known as dialysis. The desired molecules are eluted with specific substances after washing the non-interacting molecules away. Thus, this results in a highly purified material. Highly specific elution of the desired macromolecule from the stationary phase is usually effected by adding to the eluting buffer a gradient of the same kind on the macromolecule and displaces it. Possibly the most common use of affinity chromatography is for the purification of recombinant proteins. Affinity chromatography is an excellent choice for the first step in purifying a protein or nucleic acid from a crude mixture.

If the molecular weight, hydrophobicity, charge, etc. of a protein is unknown, affinity chromatography can still apply to this situation. An example of this situation is when trying to find an enzyme with a particular activity, where it can be possible to build an affinity column with an attached ligand that is similar or identical to the substrate of choice. The way that the desired enzyme would be eluted would be from the mixture based on the strong interaction of enzyme and the immobilized substrate analog, which would be done selectively through the affinity column. Then, the elution of the enzyme with the appropriate substrate can be done.

Batch and Column Setups

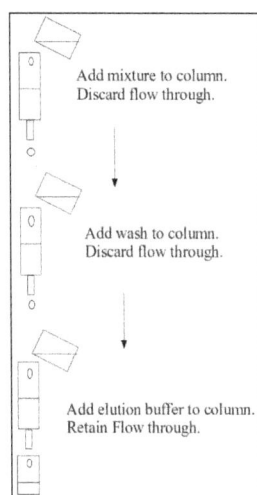

Column chromatography.

Binding to the solid phase may be achieved by column chromatography whereby the solid medium is packed onto a column, the initial mixture run through the column to allow settling, a wash buffer run through the column and the elution buffer subsequently applied to the column and collected. These steps are usually done at ambient pressure. Alternatively, binding may be achieved using a batch treatment, for example, by adding the initial mixture to the solid phase in a vessel, mixing, separating the solid phase, removing the liquid phase, washing, re-centrifuging, adding the elution buffer, re-centrifuging and removing the elute.

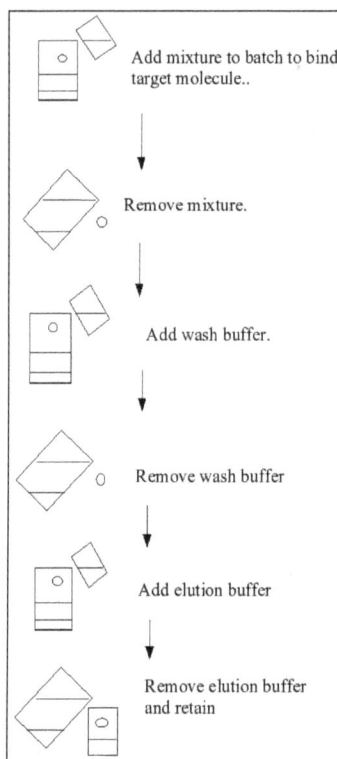

Batch chromatography.

Sometimes a hybrid method is employed such that the binding is done by the batch method, but the solid phase with the target molecule bound is packed onto a column and washing and elution are done on the column.

The ligands used in affinity chromatography are obtained from both organic and inorganic sources. Examples of biological sources are serum proteins, lectins and antibodies. Inorganic sources as moronic acts, metal chelates and triazine dyes.

A third method, expanded bed absorption, which combines the advantages of the two methods mentioned above, has also been developed. The solid phase particles are placed in a column where liquid phase is pumped in from the bottom and exits at the top. The gravity of the particles ensure that the solid phase does not exit the column with the liquid phase.

Affinity columns can be eluted by changing salt concentrations, pH, pI, charge and ionic strength directly or through a gradient to resolve the particles of interest.

More recently, setups employing more than one column in series have been developed. The advantage compared to single column setups is that the resin material can be fully loaded, since non-binding product is directly passed on to a consecutive column with fresh column material. These chromatographic processes are known as periodic counter-current chromatography (PCC). The resin costs per amount of produced product can thus be drastically reduced. Since one column can always be eluted and regenerated while the other column is loaded, already two columns are sufficient to make full use of the advantages. Additional columns can give additional flexibility for elution and regeneration times, at the cost of additional equipment and resin costs.

Specific Uses

Affinity chromatography can be used in a number of applications, including nucleic acid purification, protein purification from cell free extracts, and purification from blood.

By using affinity chromatography, one can separate proteins that bind a certain fragment from proteins that do not bind that specific fragment. Because this technique of purification relies on the biological properties of the protein needed, it is a useful technique and proteins can be purified many folds in one step.

Various Affinity Media

Many different affinity media exist for a variety of possible uses. Briefly, they are (generalized) activated/functionalized that work as a functional spacer, support matrix, and eliminates handling of toxic reagents.

Amino acid media is used with a variety of serum proteins, proteins, peptides, and enzymes, as well as rRNA and dsDNA. Avidin biotin media is used in the purification process of biotin/avidin and their derivatives.

Carbohydrate bonding is most often used with glycoproteins or any other carbohydrate-containing substance; carbohydrate is used with lectins, glycoproteins, or any other carbohydrate metabolite protein. Dye ligand media is nonspecific, but mimics biological substrates and proteins. Glutathione is useful for separation of GST tagged recombinant proteins. Heparin is a generalized affinity ligand, and it is most useful for separation of plasma coagulation proteins, along with nucleic acid enzymes and lipases

Hydrophobic interaction media are most commonly used to target free carboxyl groups and proteins.

Immunoaffinity media utilizes antigens' and antibodies' high specificity to separate; immobilized metal affinity chromatography is detailed further below and uses interactions between metal ions and proteins (usually specially tagged) to separate; nucleotide/coenzyme that works to separate dehydrogenases, kinases, and transaminases.

Nucleic acids function to trap mRNA, DNA, rRNA, and other nucleic acids/oligonucleotides. Protein A/G method is used to purify immunoglobulins.

Speciality media are designed for a specific class or type of protein/co enzyme; this type of media will only work to separate a specific protein or coenzyme.

Immunoaffinity

Another use for the procedure is the affinity purification of antibodies from blood serum. If the serum is known to contain antibodies against a specific antigen (for example if the serum comes from an organism immunized against the antigen concerned) then it can be used for the affinity purification of that antigen. This is also known as Immunoaffinity Chromatography. For example, if an organism is immunised against a GST-fusion protein it will produce antibodies against the fusion-protein, and possibly antibodies against the GST tag as well. The protein can then be covalently coupled to a solid support such as agarose and used as an affinity ligand in purifications of antibody from immune serum.

For thoroughness the GST protein and the GST-fusion protein can each be coupled separately. The serum is initially allowed to bind to the GST affinity matrix. This will remove antibodies against the GST part of the fusion protein. The serum is then separated from the solid support and allowed to bind to the GST-fusion protein matrix. This allows any antibodies that recognize the antigen to be captured on the solid support. Elution of the antibodies of interest is most often achieved using a low pH buffer such as glycine pH 2.8. The eluate is collected into a neutral tris or phosphate buffer, to neutralize the low pH elution buffer and halt any degradation of the antibody's activity. This is a nice example as affinity purification is used to purify the initial GST-fusion protein, to remove the undesirable anti-GST antibodies from the serum and to purify the target antibody.

Monoclonal antibodies can also be selected to bind proteins with great specificity, where protein is released under fairly gentle conditions.

A simplified strategy is often employed to purify antibodies generated against peptide antigens. When the peptide antigens are produced synthetically, a terminal cysteine residue is added at either the N- or C-terminus of the peptide. This cysteine residue contains a sulfhydryl functional group which allows the peptide to be easily conjugated to a carrier protein (e.g. Keyhole limpet hemocyanin (KLH)). The same cysteine-containing peptide is also immobilized onto an agarose resin through the cysteine residue and is then used to purify the antibody.

Most monoclonal antibodies have been purified using affinity chromatography based on immuno-globulin-specific Protein A or Protein G, derived from bacteria.

Immobilized Metal Ion Affinity Chromatography

Immobilized metal ion affinity chromatography (IMAC) is based on the specific coordinate covalent bond of amino acids, particularly histidine, to metals. This technique works by allowing proteins with an affinity for metal ions to be retained in a column containing immobilized metal ions, such as cobalt, nickel, copper for the purification of histidine-containing proteins or peptides, iron, zinc or gallium for the purification of phosphorylated proteins or peptides. Many naturally occurring proteins do not have an affinity for metal ions, therefore recombinant DNA technology can be used to introduce such a protein tag into the relevant gene. Methods used to elute the protein of interest include changing the pH, or adding a competitive molecule, such as imidazole.

Recombinant Proteins

Possibly the most common use of affinity chromatography is for the purification of recombinant proteins. Proteins with a known affinity are protein tagged in order to aid their purification. The protein may have been genetically modified so as to allow it to be selected for affinity binding; this is known as a fusion protein. Tags include glutathione-S-transferase (GST), hexahistidine (His), and maltose binding protein (MBP). Histidine tags have an affinity for nickel or cobalt ions which have been immobilized by forming coordinate covalent bonds with a chelator incorporated in the stationary phase. For elution, an excess amount of a compound able to act as a metal ion ligand, such as imidazole, is used. GST has an affinity for glutathione which is commercially available immobilized as glutathione agarose. During elution, excess glutathione is used to displace the tagged protein.

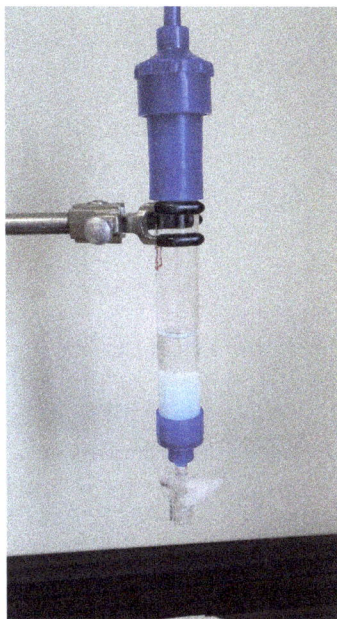

A chromatography column containing nickel-agarose beads used
for purification of proteins with histidine tags.

Lectins

Lectin affinity chromatography is a form of affinity chromatography where lectins are used to separate components within the sample. Lectins, such as concanavalin A are proteins which can bind specific alpha-D-mannose and alpha-D-glucose carbohydrate molecules. Some common carbohydrate molecules that is used in lectin affinity chromatography are Con A-Sepharose and WGA-agarose. Another example of a lectin is wheat germ agglutinin which binds D-N-acetyl-glucosamine. The most common application is to separate glycoproteins from non-glycosylated proteins, or one glycoform from another glycoform. Although there are various ways to perform lectin affinity chromatography, the goal is extract a sugar ligand of the desired protein.

Specialty

Another use for affinity chromatography is the purification of specific proteins using a gel matrix that is unique to a specific protein. For example, the purification of E. coli β-galactosidase is accomplished by affinity chromatography using p-aminobenyl-1-thio-β-D-galactopyranosyl agarose as the affinity matrix. p-aminobenyl-1-thio-β-D-galactopyranosyl agarose is used as the affinity matrix because it contains a galactopyranosyl group, which serves as a good substrate analog for

E.Coli-B-Galactosidase. This property allows the enzyme to bind to the stationary phase of the affinity matrix and is eluted by adding increasing concentrations of salt to the column.

Alkaline Phosphatase

Alkaline phosphatase from E. coli can be purified using a DEAE-Cellulose matrix. A. phosphatase has a slight negative charge, allowing it to weakly bind to the positively charged amine groups in the matrix. The enzyme can then be eluted out by adding buffer with higher salt concentrations.

Boronate Affinity Chromatography

Boronate affinity chromatography consists of using boronic acid or boronates to elute and quantify amounts of glycoproteins. Clinical adaptations have applied this type of chromatography for use in determining long term assessment of diabetic patients through analysis of their glycated hemoglobin.

Serum Albumin Purification

Of many uses of affinity chromatography, one use of it is seen in affinity purification of albumin and macroglobulin contamination. This type of purification is helpful in removing excess albumin and α_2-macroglobulin contamination, when performing mass spectrometry. In affinity purification of serum albumin, the stationary used for collecting or attracting serum proteins can be Cibacron Blue-Sepharose. Then the serum proteins can be eluted from the adsorbent with a buffer containing thiocyanate (SCN^-).

Weak Affinity Chromatography

Weak affinity chromatography (WAC) is an affinity chromatography technique for affinity screening in drug development. WAC is an affinity-based liquid chromatographic technique that separates chemical compounds based on their different weak affinities to an immobilized target. The higher affinity a compound has towards the target, the longer it remains in the separation unit, and this will be expressed as a longer retention time. The affinity measure and ranking of affinity can be achieved by processing the obtained retention times of analyzed compounds.

The WAC technology is demonstrated against a number of different protein targets – proteases, kinases, chaperones and protein–protein interaction (PPI) targets. WAC has been shown to be more effective than established methods for fragment based screening.

References

- Mcmurry, John (2011). Organic chemistry: with biological applications (2nd ed.). Belmont, CA: Brooks/Cole. P. 395. ISBN 9780495391470

- Akul Mehta (December 27, 2012). "Principle of Reversed-Phase Chromatography HPLC/UPLC (with Animation)". Pharmaxchange. Retrieved 10 January 2013

- Ito, Yoichiro (2005). "Golden rules and pitfalls in selecting optimum conditions for high-speed counter-current chromatography". Journal of Chromatography A. 1065 (2): 145–168. Doi:10.1016/j.chroma.2004.12.044

- Ninfa, Alexander; Ballou, David; Benore, Marilee (26 May 2009). Fundamental Laboratory Approaches for Biochemistry and Biotechnology. Wiley. Pp. 143–145. ISBN 978-0470087664

- Ian A. Sutherland (2007). "Recent progress on the industrial scale-up of counter-current chromatography". Journal of Chromatography A. 1151 (1–2): 6–13. Doi:10.1016/j.chroma.2007.01.143. PMID 17386930

- "Handbook for Monitoring Industrial wastewater". Environmental Protection Agency (USA). August 1973. Retrieved 30 July 2016

- Ion-chromatography, earth-and-planetary-sciences: sciencedirect.com, Retrieved 9 April, 2019

- Ninfa, Alexander J.; Ballou, David P.; Benore, Marilee (2009). Fundamental Laboratory Approaches for Biochemistry and Biotechnology (2nd ed.). Wiley. P. 133. ISBN 9780470087664

Uses of Chromatography

7

- **Applications of Food Industry**
- **Applications of Forensics**
- **Applications of Chromatography in Blood Processing**

Chromatography is applied in various areas such as forensic testing, food regulation, athlete testing, etc. Chromatographic techniques are also used as effective methods of blood purification. The topics elaborated in this chapter will help in gaining a better perspective about these uses of chromatography.

Forensic Testing

Gas chromatography is often used to investigate criminal cases.

This can take the form of crime scene testing (the analysis of blood or cloth samples), arson verification (identifying the chemicals responsible for a fire to see whether there was foul play) or blood testing after death to determine levels of alcohol, drugs or poisonous substances in the body.

Performance enhancing Drug Testing

The precision and accuracy with which chromatography can identify substances in the bloodstream make it valuable in testing for doping or performance enhancing drugs in athletes, too.

Interestingly, the news story, doping tests also work on horses, reveals how a new hybrid form of liquid chromatography combined with mass spectrometry can also be applied to our equine friends.

Ebola Immunization

As well as specialising in more flippant matters such as the quality of alcohol, chromatography may also be critical in saving millions of lives. The deadly Ebola virus, which has claimed over 5,000 lives since its outbreak late last year, has caused panic in the media and in the countries of Sierra Leone, Guinea and Liberia, to which it has been largely confined.

As scientists attempt to combat the disease, chromatography has revealed itself as incredibly useful in determining which antibodies are more effective in neutralising Ebola. Although no drugs have been conclusively validated as yet, it was instrumental in the development of the experimental immunisation Zmapp and will continue to be used in ongoing research.

Chromatographic methods will separate ionic species, inorganic or organic, and molecular species ranging in size from the lightest and smallest, helium and hydrogen, to particulate matter such as single cells. No single configuration will accomplish this, however. Little pre-knowledge of the constituents of a mixture is required. At its best, chromatography will separate several hundreds of components of unknown identity and unknown concentrations, leaving the components unchanged. Amounts in the parts per billion range can be detected with some detectors. The solutes can range from polar to nonpolar—i.e., water-soluble to hydrocarbon-soluble.

Substances of low critical temperature or low molecular weight, such as the gases at laboratory conditions showing dispersive or London intermolecular forces only, are separated with molecular sieves or gas-solid techniques. Gas-liquid chromatography is applicable to species with high critical temperatures and normal boiling points as high as 400 °C.

Substances that are solids at normal laboratory conditions with molecular weights below 1,000 are best separated with liquid-solid or liquid-liquid systems. Lower members of the molecular weight scale range are amenable to supercritical-fluid separations. Size-exclusion methods are involved at molecular weights above 1,000. Field-flow fractionation extends the size range to colloids and microscopic particles.

Separations are fast, ranging from analysis times of a few minutes to several hours. The prechromatographic world would have considered a time of several hours to separate multicomponent mixtures to be miraculously fast. Now several hours is considered excessive, and there is much emphasis on increasing speed.

Everyday uses of Chromatography

Creating Vaccinations

Chromatography is useful in determining which antibodies fight various diseases and viruses. Scientists used Chromatography in the fight against the Ebola virus, responsible for over 11,000 deaths, to develop the experimental immunisation Zmapp. The process was used to find out which antibodies are the most effective at neutralising the deadly virus.

Food Testing

The 2013 horsemeat scandal, in which horsemeat passed off by vendors as beef came to light, highlighted the ineffectiveness of traditional food analysis methods and positioned chromatography as the frontrunner in determining the contents of processed meat. The traditional methods of analysis were effective at determining the composition of raw samples but inconclusive when analysing processed meats so a more precise method was called for. High performance liquid chromatography, combined with mass spectrometry (HPLC-MS) was then successfully used to find out if meat that had been labelled as beef was beef, horse or a mix of these and various other ingredients.

Beverage Testing

Food isn't the only thing you consume which has been tested using chromatography. Many drinks manufacturers use this technique to ensure each bottle of their product is exactly the same, so you can rely on a consistent taste. One such brand is Jägermeister which uses chromatography to monitor the levels of sugar in their final product.

Drug Testing

As chromatography can accurately identify substances within the bloodstream, it is widely used in sport to test athletes for doping or performance enhancing drugs, something to think about the next time you're watching your favourite sport.

Forensic Testing

Chromatography is also used to help catch criminals. In line with programmes like CSI, gas chromatography is used to analyse blood and cloth samples, helping to identify criminals and bring them to justice.

Applications of Food Industry

Chromatography can be used in flavor studies and to detect spoilage in foods. Determining the amount of organic acids in foods provides key information about the quality of foods. Column chromatography is used to detect and quantify spoilage indicators such as pyruvic acid in milk. Pyruvic acid content is a measure of psychrotrophic bacteria present in milk.

The same separation method is used to assess total organic acid profile of milk and to measure lactose, which indicates the level of sweetness. Chromatography enables rapid analysis when compared with techniques such as bacterial plating, which may take several days to yield results. Rapid analysis is crucial in the food industry to prevent outbreak of spoilage and to minimize possible health risks.

Additive Detection

Additives are added to foods to enhance their flavors or to give them a visual appeal. For example, the presence of added malic acid in apple juice is more difficult to detect because apple juice naturally contains malic acid. However, synthetic malic acid contains fumaric acid as a contaminant and hence its level in an apple juice sample is an indicator of the commercial malic acid. Chromatography has been successfully used to detect and quantify fumaric acid in apple juice.

Determining Nutritional Quality

Vitamin C depletion in foods can be an indicator of depletion of other nutrients and so the vitamin C content of foods and beverages is closely monitored during all stages of food processing using column chromatography. This analysis can be carried out rapidly using modern acid analysis columns coupled with electrochemical detection even in complex samples. This technique is used to quantitate vitamin C in juices, powdered drinks, and both fresh and frozen vegetables and fruits.

Applications of Forensics

Gas chromatography is used to test evidence such as blood or hair from a crime scene. This allows investigators to understand the crime better and to develop theories on what exactly happened and where the victim has been earlier, based on the material found.

Forensic Pathology

Gas chromatography (GC) has been widely used in forensic pathology to identify the type of compounds and fluids present in the human body, post death. This testing can help detect the presence of alcohol or drugs or poisonous substances in the body at the time of death, thus assisting in determining the possible motive and cause of death.

Arson Investigation

GC is a low cost technique used to identify ignitable/flammable liquids from fire debris. On comparison with a list of flammable liquids publically available, the exact kind of liquid used can be concluded. Mass spectrometry (MS) characterization of the separated components yields better and more precise results.

Molecular Biology Studies

Hybrid techniques that combine electrochemistry (EC) and MS with chromatography are powerful tools in the study of redox reactions involving various bioorganic molecules. ESI-MS is coupled with liquid chromatography (LC) separation for the characterization of the reaction mixture. EC–LC–MS is applied in the study of biomolecules such as proteins, peptides, and nucleic acids.

Metabolomics and Proteomics

EC–LC–MS is essential in mimicking biotransformation reactions, such as phase I oxidative reactions in drug metabolism studies. The technology has been applied in the study of pharmaceutical compounds such as; acetaminophen, diclofenac, lidocaine, clozapine, haloperidol, flunitrazepam, chlorpromazine, alprenolol, albendazole and verapamil.

In proteomics, this technique is used to analyze oxidation of proteins and peptides and in selective labeling of these substances. Chromatography techniques are also widely used in purification of plasma proteins, hormones, monoclonal antibodies, and vaccines as part of their development.

Nucleic Acids Research

Electrochemistry coupled with LC, MS, or gas chromatography (GC) has been successfully used to identify the oxidation products of nucleobases, nucleotides, and nucleosides. This has accelerated the identification of these compounds compared to long drawn-out isolation steps.

Applications of Chromatography in Blood Processing

Chromatographic techniques have been used in blood processing and purification since the 1980s. It has emerged as an effective method of purifying blood components for therapeutic use.

Human Blood Plasma

Blood plasma is the liquid component of blood, which contains dissolved proteins, nutrients, ions, and other soluble components. In whole blood, red blood cells, white blood cells, and platelets are suspended within the plasma. The goal of plasma purification and processing is to extract specific materials that are present in blood, and use them for restoration and repair. There are several components that make up blood plasma, one of which is the protein albumin. Albumin is a highly water-soluble protein with considerable structural stability. It serves as a transportation device for materials such as hormones, enzymes, fatty acids, metal ions, and medicinal products. It is also used for therapeutic purposes, being essential in restoration and maintenance of circulating blood volume in imperative situations such as severe trauma or surgery. With little room for error, extremely pure samples that are lacking impurities needs to be at hand in good amount. Human blood plasma is important for the body so the nutrients etc. can be stored.

Development of Chromatography

Traditionally, the Cohn process incorporating cold ethanol fractionation has been used for albumin purification. However, chromatographic methods for separation started being adopted in the early 1980s. Developments were ongoing in the time period between when Cohn fractionation started being used, in 1946, and when chromatography started being used, in 1983. In 1962, the Kistler & Nistchmann process was created which was a spinoff of the Cohn process. Chromatographic processes began to take shape in 1983. In the 1990s, the Zenalb and the CSL Albumex processes were created which incorporated chromatography with a few variations.

The general approach to using chromatography for plasma fractionation for albumin is: recovery of supernatant I, delipidation, anion exchange chromatography, cation exchange chromatography, and gel filtration chromatography.

The recovered purified material is formulated with combinations of sodium octanoate and sodium N-acetyl tryptophanate and then subjected to viral inactivation procedures, including pasteurisation at 60 °C.

This is a more efficient alternative than the Cohn process for four main reasons: 1) smooth automation and a relatively inexpensive plant was needed, 2) easier to sterilize equipment and maintain a good manufacturing environment, 3) chromatographic processes are less damaging to the albumin protein, and 4) a more successful albumin end result can be achieved.

Compared with the Cohn process, the albumin purity went up from about 95% to 98% using chromatography, and the yield increased from about 65% to 85%. Small percentage increases make a difference in regard to sensitive measurements like purity. There is one big drawback in using chromatography, which has to do with the economics of the process. Although the method was efficient from the processing aspect, acquiring the necessary equipment is a big task. Large machinery is necessary, and for a long time the lack of equipment availability was not conducive to its widespread use. The components are more readily available now but it is still a work in progress and will possibly be ready in the future to help the world.

Bridging Methods

Integrating traditional and modern methods is a useful way to process albumin.

There are three main steps that combine Cohn fractionation with chromatography: 1) factors I, II, and III are removed via cold ethanol fractionation, 2) Sepharose fast flow ion exchange and sepharose fast flow chromatography procedures are run, and 3) gel filtration is run. The result is albumin with 9% lower aluminum levels with a processing time that is almost twice as fast.

Although it was hard to make chromatographic processing methods widely adopted, global expansion is a work in progress. Various blood components must be readily available at various medical treatment centers around the world. The Institute of Transfusion Medicine in Skopje, Macedonia is a plasma fractionation center in the Balkans. Their modernized albumin purification process consists of five steps:

- Starting material is plasma that has been pretreated by centrifugation,

- A round of gel filtration is run,

- Ion exchange on DEAE Sepharose is run to bind the albumin to the column,

- Albumin is eluted with a sodium acetate buffer, and

- Final polishing with gel filtration.

The end result is a highly pure and safe batch of albumin that is 100% non-pyrogenic, sterile, and free of active HIV virus. The product purity is greater than 98% and the protein content is about 50 g/L.

Non-chromatographic Processing Methods

Other plasma processing methods exist, but generally do not provide the resolution or purity of chromatographic methods. Two-phase liquid extraction may be performed using polyethylene glycol (PEG)-phosphate Aqueous two-phase systems, with a PEG-rich top layer and a phosphate-rich bottom layer. Although this method is somewhat useful for protein recovery, it does not work as well for the recovery of other blood components. Membrane fractionation has the advantage of minimal protein loss yet high removal of pathological plasma components. This method incorporates processes such as thermofiltration and applying pulsate flow. The latest two-stage membrane system utilizes a high flow recirculation circuit that is effective for removal of LDL cholesterol. It may prove useful for patients that have clogged arteries and other cardiovascular problems involving cholesterol. Batch adsorption, e.g. onto ion exchange media, is only useful when dealing with smaller samples of plasma, typically 200 mL or less. Batch adsorption recovers the product in a larger volume of elution buffer than does column chromatography or frontal chromatography, and the resulting more dilute product requires concentration, typically on a membrane system, which can lead to loss of product by irreversible adsorption to the membrane.

References

- 5-uses-of-chromatography-in-everyday-life, breaking-news, 39, industrial-news, news: chromatographytoday.com, Retrieved 25 February, 2019

- Plate-height#ref80531, chromatography, science: britannica.com, Retrieved 26 July, 2019

- 5-everyday-uses-for-chromatography: peakscientific.com, Retrieved 8 January, 2019

- Life-Science-Applications-of-Chromatography, life-sciences: news-medical.net, Retrieved 6 June, 2019

- Matejtschuk P, Dash CH, Gascoigne EW (December 2000). "Production of human albumin solution: a continually developing colloid". Br J Anaesth. 85 (6): 887–95. Doi:10.1093/bja/85.6.887. PMID 11732525

PERMISSIONS

We would like to thank the editorial team for lending their expertise to make the book truly unique. They have played a crucial role in the development of this book. Without their invaluable contributions this book wouldn't have been possible. They have made vital efforts to compile up to date information on the varied aspects of this subject to make this book a valuable addition to the collection of many professionals and students.

This book was conceptualized with the vision of imparting up-to-date and integrated information in this field. To ensure the same, a matchless editorial board was set up. Every individual on the board went through rigorous rounds of assessment to prove their worth. After which they invested a large part of their time researching and compiling the most relevant data for our readers.

The editorial board has been involved in producing this book since its inception. They have spent rigorous hours researching and exploring the diverse topics which have resulted in the successful publishing of this book. They have passed on their knowledge of decades through this book. To expedite this challenging task, the publisher supported the team at every step. A small team of assistant editors was also appointed to further simplify the editing procedure and attain best results for the readers.

Apart from the editorial board, the designing team has also invested a significant amount of their time in understanding the subject and creating the most relevant covers. They scrutinized every image to scout for the most suitable representation of the subject and create an appropriate cover for the book.

The publishing team has been an ardent support to the editorial, designing and production team. Their endless efforts to recruit the best for this project, has resulted in the accomplishment of this book. They are a veteran in the field of academics and their pool of knowledge is as vast as their experience in printing. Their expertise and guidance has proved useful at every step. Their uncompromising quality standards have made this book an exceptional effort. Their encouragement from time to time has been an inspiration for everyone.

The publisher and the editorial board hope that this book will prove to be a valuable piece of knowledge for students, practitioners and scholars across the globe.

INDEX

www.ingramcontent.com/pod-product-compliance
Lightning Source LLC
Chambersburg PA
CBHW061246190326
41458CB00011B/3595